Praise for Hothouse Earth

'The Earth is already in a dangerous phase of heating. Many scientists admit privately to actually being "scared" by recent weather extremes. But the public doesn't like pessimism, so we environment journalists hint at future optimism. This book provides a more steely-eyed view on how we can cope with a hothouse world.'

Roger Harrabin, former BBC Environment Analyst

'Taut, calmly told and truly terrifying – and there's no arguing with the science. If you read just one book about the menace of climate breakdown, make it this one.'

Tim Radford, Climate News Network

'A comprehensive tour of climate breakdown, the trouble we are heading for and the many forms it might take. This accessible and authoritative book is a must-read for anyone who still thinks it could be OK to carry on as we are for a little bit longer, or that climate chaos might not affect them or their kids too badly.'

Mike Berners-Lee, professor at Lancaster University, founder of Small World Consultancy and author of *There is No Planet B: A Handbook for the Make or Break Years*

'Professor Bill McGuire has a rare talent for presenting authoritative and complex information in writing that is both accessible and enjoyably fluid. His book is convincing and passionate – an invaluable guide for those who are relatively new to the issue of climate breakdown and a useful revisor for those of us who have been reading the science for many years.'

Brendan Montague, editor of the *Ecologist*

'A compelling clarion call for a planet in peril. If the searing science of *Hothouse Earth* doesn't set alarm bells ringing, then it is difficult to see what else will.'

Professor Iain Stewart, geologist and broadcaster

'Ironically, it's never been harder telling the full truth about the climate emergency. That truth is so shocking. So painful. It invites rejection. But there can be no authentic hope for a better world without that truth being unflinchingly spelled out. Thanks then to

Bill McGuire for doing exactly that in *Hothouse Earth* – and for still leaving us with plenty of reasons to be hopeful – just so long as we get our shit sorted without any further delay.'

Jonathon Porritt, environmental campaigner and author of *Hope in Hell: A Decade to Confront the Climate Emergency*

'It is rare indeed for a top scientist to spell out with blunt honesty the hell that we are heading into. Bill McGuire is one of the very few.'

Roger Hallam, co-founder of Extinction Rebellion and co-author of *This is Not a Drill: An Extinction Rebellion*

'There's a climate emergency on and our leaders haven't understood just how serious it is. In this concise book, Professor Bill McGuire expertly lays out the scale of the threat in very clear terms – including how much damage we have already done. He points out just how little time we have left to stop the climate crisis engulfing human civilisation. Every decision maker in government, business and wider society should read this book – and then act as fast as possible to reduce carbon pollution to zero.'

Dr Stuart Parkinson, executive director, Scientists for Global Responsibility

'*Hothouse Earth* might accurately be described as a bit of a grim read, but there is no hyperbole here. Everything in Prof. McGuire's book is solidly based upon peer-reviewed research and current observations. McGuire wants us to face up to the facts; when one does so, then one has no choice any longer. Then, grim is the point; and actually, grim is the way. For, *only if we get people to feel less shy and embarrassed about talking and hearing grim do we stand a chance* ... For, only if we are ready to be real about our predicament have we any hope of measuring up to it. If you are after light reading, or just want to put on a happy face, don't buy this book. Only those ready for a strict diet of truth should dare open it. *Hothouse Earth* is an easy to understand and authoritative reference source for all things climate science. It is a very, very, sobering read. If our so-called leaders were to read it, they would adapt. They would change (or else we must change them for others up to the job). Why not buy them a copy?'

Rupert Read, Associate Professor of Philosophy at the University of East Anglia and former spokesperson and strategist for Extinction Rebellion. His new book, *Why Climate Breakdown Matters* is published in August 2022

Hot Science is a series exploring the cutting edge of science and technology. With topics from big data to rewilding, dark matter to gene editing, these are books for popular science readers who like to go that little bit deeper ...

AVAILABLE NOW AND COMING SOON:

Dark Matter & Dark Energy:
The Hidden 95% of the Universe

Outbreaks & Epidemics:
Battling Infection From Measles to Coronavirus

Rewilding:
The Radical New Science of Ecological Recovery

Hacking the Code of Life:
How Gene Editing Will Rewrite Our Futures

Origins of the Universe:
The Cosmic Microwave Background
and the Search for Quantum Gravity

Behavioural Economics:
Psychology, Neuroscience,
and the Human Side of Economics

Quantum Computing:
The Transformative Technology of the Qubit Revolution

The Space Business:
From Hotels in Orbit to Mining the Moon
– How Private Enterprise is Transforming Space

Game Theory:
Understanding the Mathematics of Life

Nuclear Fusion:
The Race to Build a Mini-Sun on Earth

Hot Science series editor: Brian Clegg

HOTHOUSE EARTH
AN INHABITANT'S GUIDE

BILL McGUIRE

ICON

Published in the UK and USA in 2022
by Icon Books Ltd, Omnibus Business Centre,
39–41 North Road, London N7 9DP
email: info@iconbooks.com
www.iconbooks.com

Sold in the UK, Europe and Asia
by Faber & Faber Ltd, Bloomsbury House,
74–77 Great Russell Street,
London WC1B 3DA or their agents

Distributed in the UK, Europe and Asia
by Grantham Book Services,
Trent Road, Grantham NG31 7XQ

Distributed in the USA
by Publishers Group West,
1700 Fourth Street, Berkeley, CA 94710

Distributed in Australia and New Zealand
by Allen & Unwin Pty Ltd,
PO Box 8500, 83 Alexander Street,
Crows Nest, NSW 2065

Distributed in South Africa
by Jonathan Ball, Office B4, The District,
41 Sir Lowry Road, Woodstock 7925

Distributed in India by Penguin Books India,
7th Floor, Infinity Tower – C, DLF Cyber City,
Gurgaon 122002, Haryana

Distributed in Canada by Publishers Group Canada,
76 Stafford Street, Unit 300,
Toronto, Ontario M6J 2S1

ISBN: 978-178578-920-5

Typeset in Iowan by Marie Doherty

Printed and bound in Great Britain
by Clays Ltd, Elcograf S.p.A.

This book is dedicated to valiant climate activists everywhere, who are daily fighting ignorance, abuse, obfuscation and outright denial. I know you will win the battle, because you simply can't afford to lose.

ABOUT THE AUTHOR

Bill McGuire is Professor Emeritus of Geophysical and Climate Hazards at University College London, a co-director of the New Weather Institute and was a contributor to the 2012 IPCC report on climate change and extreme events. His books include *A Guide to the End of the World: Everything You Never Wanted to Know* and *Waking the Giant: How a Changing Climate Triggers Earthquakes, Tsunamis and Volcanoes*. His first novel, *Skyseed* – an eco-thriller about climate engineering gone wrong – was published in 2020. He writes for many publications, including the *Guardian*, *The Times*, the *Observer*, *New Scientist*, *Science Focus* and *Prospect* and is author of the Cool Earth blog on Substack. Bill lives, runs (sometimes) and grows fruit and veg in the wonderful English Peak District, where he resides with his wife Anna, sons Jake and Fraser and cats Dave, Toby and Cashew.

CONTENTS

CONVERSION TABLE FOR KEY TEMPERATURES

Temperature increase values		Absolute temperature values	
Centigrade	Fahrenheit	Centigrade	Fahrenheit
0.5	0.9	20	68
1	1.8	25	77
1.5	2.7	30	86
2	3.6	35	95
2.5	4.5	40	104
3	5.4	45	113
3.5	6.3	50	122

FOREWORD

This book was written mostly over a six-month period, during the course of which the COP26 (the 26th Conference of Parties) UN Climate Change Conference was staged in Glasgow. Putting it together has been quite a challenge, not only because there is such a vast amount of material out there, but also because both the science and the policy are constantly changing. In retrospect, having to squeeze a quart into a pint pot has actually worked in my favour, and hopefully yours too, by helping to concentrate the mind and forcing me to zero in on the core issues at the heart of the climate emergency. The end product is a small book with a big message.

It is fortuitous that the final pulling together of material coincided with the COP26 summit, providing – as it did – a more credible idea of where we are likely headed. It was billed by many, including me, as arguably the most important meeting in the history of humankind, and I attended with grateful humility and with this always in mind. Hopes were high that the outcome of the two weeks of discussion, debate and negotiation might be a glimpse of a realistic pathway out of the dark place we find ourselves in; an attainable route towards the goal of keeping the global average temperature rise (since pre-industrial times) this side of the 1.5°C (2.7°F) so-called dangerous climate change guardrail. Unfortunately, this was not to be.

True, there were plenty of fine promises, on everything from protecting forests to reducing methane emissions, cutting out coal and throwing cash at the majority, or developing, world to help fund carbon-reducing measures, but on the detailed mechanisms, legal frameworks and monitoring required to ensure fulfilment of these pledges there was next to nothing.

Some early post-COP26 modelling averred that, *if* pledges were all met and targets achieved, then we might be on track for 'just' a 1.9°C (3.4°F), or even 1.8°C (3.2°F), global average temperature rise. Firstly, however, this is a very big *if* indeed. Secondly, such predictions fly in the face of peer-reviewed research published pre-COP26, which argues that a rise of more than 2°C (3.6°F) is already 'baked-in' or, in plain language, certain.

The post-COP26 reality is this. To have even the tiniest chance of keeping the global average temperature rise below 1.5°C, we need to see emissions down 45 per cent by 2030. In theory, this might be possible, but in the real world – barring some unforeseen miracle – it isn't going to happen. Instead, we are on course for close to a 14 per cent rise by this date that will almost certainly see us shatter the 1.5°C guardrail in less than a decade.

This book takes as its starting premise, then, the notion that, practically, there is now no chance of dodging a grim future of perilous, all-pervasive, climate breakdown. It is no longer a matter of what we can do to avoid it, but of what we should expect in the decades to come, how we can adapt to a hothouse world with more extreme weather and what we can do to stop a bleak situation deteriorating even further.

I ought to make clear here that the terms 'hothouse Earth' or 'greenhouse Earth' are used formally, in a definitive sense, to describe the state of our planet in the geological past when global temperatures have been so high that the poles have been ice-free. A hothouse *state*, however, is not required for hothouse *conditions*, which are already becoming far more commonplace, and fast becoming the trademark of our broken climate. What I mean by hothouse Earth, then, is not an ice-free planet, but a world in which lethal heatwaves and temperatures in excess of 50°C (122°F) in the tropics are nothing to write home about; a world where winters at temperate latitudes have dwindled to almost nothing and baking summers are the norm; a world where the oceans have heated beyond the point of no return and the mercury climbing to 30°C+ (86°F+) within the Arctic Circle is no big deal.

To all intents and purposes, *this* is the hothouse planet we are committed to living on; one that would be utterly alien to our grandparents. The fact is that a temperature rise of 2°C – which is likely the minimum we are committed to – may not sound like much, but remember that this is an average temperature. In some parts of the planet, the increase will be far higher. In addition, even this small rise will drive extreme high temperature events beyond anything experienced in human history. A child born in 2020 will face a far more hostile world than its grandparents. Compared with someone lucky enough to be born in 1960, one study estimates that – on average – they will experience seven times more heatwaves, twice as many droughts and three times as many floods and harvest failures. The reality could very well be far worse, and it will be for the billions of vulnerable people living in the majority world. Looking at the broader picture, anyone younger than 40 today will suffer ever more frequent bouts of extreme weather that would be virtually impossible in the absence of global heating.

In the pages that follow, I have tried – using the most recent observations and latest peer-reviewed research – to shine a light on how our familiar world is already changing, and how it will be completely reshaped in the decades ahead. I have sought to do this by focusing on three facets of the climate emergency. First, to place the changes to our climate caused by human action in context, by taking a trip through our planet's 4.6-billion-year history, during which time conditions ranged from icehouse to hothouse and back again on many occasions. Second, to zero in on the current global heating episode and drill down into the evidence for the accelerating breakdown of our once stable climate. Third, to dissect the wide-ranging consequences of contemporary heating in more detail, from storm, flood, fire and drought to mass migration, water wars and health issues, along with hard-to-predict 'stings in the tail', such as Gulf Stream collapse and methane 'bombs'. A concluding section looks ahead to possible futures, addresses what we need to do now to minimise the impact of dangerous climate breakdown and considers whether technology can save us. It also rams home the message that – even at this late stage – it remains vital that we cut emissions to the bone as soon as we possibly can.

As a trained geologist, I have always tended to set more store by observation and measurement than modelling or simulation, although both certainly have their place, particularly in trying to pin down future climate scenarios. Consequently, the content of this book is driven as much by current observation of climate trends and knowledge of past episodes of climate change as it is by model-based forecasts of what we might expect in the decades and centuries to come.

This is important because there is little doubt that our climate is changing for the worse far quicker than predicted by earlier models. It is also the case, research has revealed, that climate scientists – as a tribe – tend to gravitate towards a consensus viewpoint rather than go out on a limb, and they are inclined to make forecasts that underplay the reality. This does not help us minimise the impacts of the dangerous climate breakdown that we now know is on our doorstep. Far from it. In order to be as prepared as we can be, we need to plan for the worst even as we hope for the best.

A note on terminology. 'Global warming' has a cosy feel to it that is far from justified by the reality, while the rapidly increasing incidences of extreme weather show that our once stable climate is not simply changing, but well on its way to failing. To reflect this, I will be mostly using the alternative terms 'global heating' and 'climate breakdown'. Both are coming into general usage because they far more accurately describe what is happening to our world and our climate.

GROUND ZERO

1

Cromford, England, 1771

Two hundred and fifty years ago, a man called Richard
Arkwright sparked a revolution. No violence was involved,
there was no blood. The Sun shone as if nothing had happened, but its rising and setting bookended a day that
changed the world for ever.

Arkwright, originally a lowly barber, wigmaker and
tooth-puller from Preston, Lancashire, had a big idea and put
it to work in the small Derbyshire village of Cromford, just
down the road from my home up here in the Peak District.

Having made a tidy bit of cash as a result of a waterproof
dye he had invented for wigs, Arkwright developed an interest in the cotton industry, which was burgeoning at the time,
especially in Lancashire. A canny operator, he was quick to
see that making yarn by hand, using an apparatus known as
the Spinning Jenny, could never keep up with the huge and
growing demand for the product. His big idea was to develop
and patent a mechanised spinning device known as the water

frame. Not only could this make yarn more speedily, but the product was also stronger and of better quality.

Traditionally, women spun the yarn and their menfolk wove it into cloth, all within the confines of small cottages. But Arkwright's invention changed all this. The water frame spinning machines were far too big to fit into a home, and – as the name testifies – they needed flowing water to operate.

Arkwright's solution was to construct a large building close to the River Derwent to host his spinning machines, and power them by means of water wheels. The new facility looked just like a water mill used to make flour from grain and soon became known simply as Cromford Mill. Spinners and weavers living in Cromford village, close by the new installation, saw their lives change overnight. Instead of spinning at home, the women – and children as young as seven too – now headed every day to the mill to produce yarn using the water frames, while the men stayed at home to weave the yarn into cloth.

At a stroke, Arkwright – and his partners Samuel Need and Jedediah Strutt – had unleashed mass manufacturing on an unsuspecting world. To squeeze the most out of his workers, Arkwright instigated a system of overlapping thirteen-hour shifts, with bells at 5am and 5pm to rouse the unfortunate women and children and set them commuting, bleary-eyed, down the steep hill from their homes to the mill. Nothing less than a modern, mechanised, factory system was up and running.

The mechanisation of work previously undertaken by individuals or small groups, using human muscle and dexterity alone, spread like wildfire in the years that followed. Arkwright himself built new mills along the Derwent, in Lancashire and Scotland. Soon, the arrival of steam power freed new developments from needing access to flowing water, and new factories were built for the mass production of glass, chemicals and machine tools, such as lathes and steam hammers. Machine tool production boosted the revolution in work and manufacturing even further, so that within decades a national economy previously entirely reliant upon manual labour was transformed beyond all recognition.

Steam-powered machine technology revolutionised textile production, coal refining, metalworking and a multitude of other enterprises. At the same time, an explosion of new roads, canals and railway lines broadcast the revolution into every corner of the UK and, eventually, far beyond its borders.

Arkwright's revolution, born on that sunny Derbyshire day in 1771, was nothing less than the Industrial Revolution. In the century that followed, a tsunami of mechanisation and mass production swept across Europe and North America, destroying one way of life and replacing it with another; a seemingly unstoppable wave of change that continues to roar across the face of our planet today.

Arkwright's legacy

The work of the former wigmaker is recognised at the highest level by the fact that Arkwright's original Cromford Mill, and the others built in the late 18th century upon the banks of the River Derwent, now constitute a UNESCO World Heritage Site marking the unequivocal birthplace of the Industrial Revolution. Arkwright's legacy is widely hailed as a colossal achievement, of immense benefit to humankind, and so it is in many ways, but it also comes at enormous cost.

As Arkwright's baby has grown and developed, it has become an unstoppable beast. Increasingly, and in particular since the end of the Second World War, the wigmaker's revolution has underpinned a global economy governed by a conspiracy of largely unfettered, free-market capitalism and a desperate and growing urge to consume. This has driven an immeasurable rise in the quality of life in many countries. At the same time, however, billions have remained mired in poverty, while the gap between the haves and have-nots grows ever wider. The consequences of market-driven consumerism have also been dire, with widespread pollution, the large-scale degradation and destruction of the environment and an overheating planet the result.

A straight line can be drawn, then, from the opening of Arkwright's Cromford Mill to the ongoing climate and ecological emergency, the greatest existential threat our civilisation has ever faced. As far as climate change is concerned, the small Derbyshire village of Cromford is nothing less than ground zero. Here, in other words, is where all our problems began.

The climate emergency: how did we get here?

Time was, people were content to have what they needed for a good life; now it seems – at least in the industrialised nations – that we can never have too much stuff. On our tiny planet, a society based upon ever more consumption of rapidly diminishing resources cannot be sustained, which is why we find ourselves, today, in the midst of a climate and ecological crisis.

When Arkwright opened his mill 250 years ago, the concentration of carbon dioxide in the Earth's atmosphere was

280 parts per million (ppm). Carbon dioxide is the principal greenhouse gas, which acts to keep our planet warm and helps shield it from the bitter cold of space. But too much of it can lead to a dangerous rise in global temperature. The level of carbon dioxide goes up and down over geological time in response to natural processes. A level of 280 ppm is, for example, near enough what you would expect in the current interglacial period during which human civilisation has burgeoned. At the height of the last ice age, just 20,000 years ago, levels fell to around 180 ppm, and they would do so again during the next ice age, if it isn't kept at bay by humankind's adulteration of the atmosphere.

This is because Arkwright's legacy is not only the creation of an economic wonder capable of meeting all our wants and needs, but a prodigious exhalation of pollution that has seen an additional 2.4 *trillion* tonnes of carbon dioxide added to our planet's atmosphere. This raised levels in 2021 to a high of 420 ppm – a hike of 50 per cent – bringing with it a global average temperature rise of 1.2°C and an increasingly obvious upsurge in extreme weather events, as our once stable climate starts to break down.

Perhaps the most depressing thing about the growing climate emergency is that we have been put on notice time and time again about the potentially catastrophic impact of rising greenhouse gas levels in the atmosphere, but we have repeatedly refused to listen and chosen not to act.

No one can say we weren't warned

Way back in 1856, the US scientist Eunice Foote wrote a paper describing the astonishing heat-absorbing properties of carbon dioxide. The paper was based upon the results of a simple but effective experiment, which involved placing two jars, one filled with air and the other with carbon dioxide, in full sun. The jar filled with carbon dioxide heated up much more than the air-filled one, leading Foote to conclude that carbon dioxide in the atmosphere would absorb the Sun's heat in the same way.

She even speculated that 'if the air had mixed with it a higher proportion of carbon dioxide than at present, an increased temperature would result'. This was, effectively, the first prediction of global warming, made more than 150 years ago.

A few years on, in the 1860s, the Irish scientist John Tyndall took Foote's work further. On the basis of hundreds of experiments on the properties of a range of gases, Tyndall noted that variations in the level of atmospheric carbon dioxide, along with water vapour and methane – both also greenhouse gases – 'must produce a change of climate'.

Fast forward to the end of the 19th century and we find Swedish physicist Svante Arrhenius laying the groundwork for modern ideas that link carbon dioxide levels in the atmosphere and global temperature. Arrhenius recognised that a halving of carbon dioxide levels could explain the fall in temperature associated with an ice age. He also forecast that a doubling of carbon dioxide levels would result in the global average temperature climbing by 5–6°C, a figure he later revised downwards to 4°C.

The amount by which our planet will warm for a doubling of the atmospheric carbon dioxide level (compared to Arkwright's time) is known as the 'climate sensitivity'. Even today, there is considerable debate about how high this number is, with estimates mainly ranging from 1.5°C to

4°C. Most recently, climate models have coalesced around a figure of around 3.7°C, astonishingly close to Arrhenius' prediction.

During the 20th century, it slowly became clear that not only *could* increasing levels of atmospheric carbon dioxide warm the planet, but that this was actually happening. In the late 1930s, the British engineer and amateur climate scientist Guy Callendar demonstrated that both land temperatures and carbon dioxide levels had risen over the course of the previous 50 years and proposed that the two were linked. Callendar's ideas were met with widespread scepticism, but later studies proved him to be spot-on.

In the 1950s, the Canadian physicist Gilbert Plass examined the links between increasing carbon emissions due to human activities and global temperatures more closely. In an especially prophetic interview with *Time* magazine in 1953, he is quoted as saying: 'At its present rate of increase, the CO_2 in the atmosphere will raise the earth's average temperature 1.5° Fahrenheit every 100 years … for centuries to come, if man's industrial growth continues, the earth's climate will continue to grow warmer.'

While this initial estimate of the rate of temperature increase is much too low, the quote still points to an extraordinary level of far-sightedness. Plass later went on to forecast that the value of climate sensitivity would be 3.6°C and that atmospheric carbon dioxide levels in 2000 would be 30 per cent higher than in 1900, pushing up the global average temperature by 1°C. As previously mentioned, the current estimate of climate sensitivity is 3.7°C, while atmospheric carbon levels in 2000 were actually 37 per cent up on 100 years earlier, leading to warming of around 0.7°C. These, then, are amazingly impressive predictions.

In 1956, another US scientist called Roger Revelle, working in the same field, testified before Congress that the Earth was a spaceship which – due to increasing carbon emissions – was threatened by rising sea levels and desertification. An article describing Revelle's research, published the following year, used the term 'global warming' for the first time, concluding that Revelle's findings pointed to large-scale global warming with the potential to cause radical climate changes.

After the 1950s, research linking global warming with carbon dioxide emissions and their ever-increasing accumulation in the atmosphere moved fast. Most critically, in 1961, US scientist Charles Keeling demonstrated that carbon dioxide levels were climbing incrementally, but steadily, year on year. He used his measurements to define what has become known as the Keeling Curve, a graph that continues to record the remorseless rise of atmospheric CO_2 from 315 ppm, when Keeling began his measurements, to 420 ppm in 2021.

As understanding of what was happening gathered pace, so too did the urgency of the warnings. In 1966, the US Nobel Laureate chemist Glenn Seaborg cautioned that the rate at which carbon dioxide was being added to the atmosphere (6 billion tons a year at the time) could lead to marked changes to the climate that could be beyond our control.

A couple of years on, a Stanford University report written for the American Petroleum Institute in 1968 flagged the seriousness of global warming, concluding that significant temperature changes were almost certain to occur by the year 2000 and that these could bring about climatic changes. This publication is especially notable in that it was completely ignored by the fossil fuel industry that commissioned it. Not only did it take no action whatsoever, but in the decades that followed, oil companies launched a multi-million-dollar drive of obfuscation and denial designed to sow confusion and undermine public confidence in climate science and stymie the need for urgent action to tackle global warming.

Notwithstanding frigid scare stories in the tabloids that a new ice age was on the way, support for the idea of global warming and its link to carbon emissions continued to build within the climate science community throughout the 1970s and into the 1980s. In a seminal presentation to the US Senate Committee on Energy and Natural Resources in 1988, illustrious climate scientist James Hansen, who at the time was working for NASA, warned that global warming was happening now and was without doubt a consequence of escalating greenhouse gas emissions. Hansen's testimony is widely recognised as being a pivotal moment that marked the beginning of the global warming policy debate that continues today. At a stroke, it also ensured that awareness of the threat escaped from its climate science silo and entered the public arena, wherein its profile has grown ever since.

As the broader ramifications became increasingly clear, the term 'global warming' was replaced, at least in scientific and technical circles, by 'climate change'. This updated terminology was important as it broadened the focus of debate from what was happening to our planet – a steady rise in global temperature due to carbon emissions – to incorporate the consequences of warming – the large-scale modification of our climate.

The Intergovernmental Panel on Climate Change

Unfortunately, more widespread knowledge and understanding of our changing climate has only translated very slowly, too slowly, into action. The same year as Hansen's testimony, the World Meteorological Organization established the Intergovernmental Panel on Climate Change (IPCC), with the aim of evaluating climate change science, coordinating research findings and promoting wider understanding of climate change.

To this day, the IPCC remains *the* arbiter of climate science, and – when it comes to understanding and recording what is happening to our climate – its assessment reports engender almost biblical reverence. The panel's principal operational objective is to provide governments at all levels with scientific information that they can use to develop climate policies. To say that the formulation of such policies has taken place at a snail's pace would – in all honesty – be insulting to molluscs.

When the IPCC's first assessment report was published in 1990, today's younger climate scientists and politicians were still at infant school, or even babes in arms. The report noted, with certainty, that emissions arising from human activities were substantially increasing the atmospheric concentrations of the greenhouse gases carbon dioxide, methane, chlorofluorocarbons and nitrous oxide. The authors went on to forecast significant rises in global temperature and sea level if the world continued with business as usual. Calls to action accompanied the launch of the report but fell on deaf ears.

More than 30 years on, many of those ears – and quite a few new ones – remain deaf. Business as usual continues in most countries, and global greenhouse gas emissions have maintained – barring a small Covid-related downward blip in 2020 – their year-on-year climb. In the interval between the publication of the IPCC's first and sixth assessments, the first part of the latter of which was released in 2021, total greenhouse gas emissions have risen by 43 per cent, from a little under 35 billion tonnes to more than 50 billion tonnes. Over the same period, the atmospheric concentration of carbon dioxide has shot up from 354 ppm to 420 ppm – an increase of nearly 19 per cent. Worse still, according to the US National Oceanic and Atmospheric Administration, the warming influence of the whole basket of greenhouse gases has climbed a staggering 47 per cent since the IPCC first appeared on the scene.

This is probably the best place to reinforce the point that carbon dioxide is not the only greenhouse gas on the up. Atmospheric levels of other gases have surged over the past few decades too, most notably methane, which, as a greenhouse gas, punches far above its weight when it comes to its warming potential. Methane is produced by natural means, including releases from wetlands and ocean and lake beds. It is also a product of farting livestock, rice paddies, the oil and gas industry and thawing permafrost (the enormous tracts of permanently frozen ground at high latitudes) – all linked to human activities – which explains why methane levels have climbed two and a half times since Arkwright's time.

Over a period of twenty years, a tonne of methane causes 86 times as much atmospheric warming as a tonne of carbon dioxide. It is lucky, then, that it hangs around for much less time, so that only 250 kilograms would be left in the atmosphere after twenty years. Nonetheless, close to one-third of the warming since pre-industrial times can be attributed to rising concentrations of the gas. It is alarming to realise that when the warming effect of all greenhouse gases is taken together, it is equivalent, today, to an atmosphere containing 500 ppm carbon dioxide.

If world leaders had taken purposeful avoiding action in 1990, when the IPCC launched its first report, we could well be on top of the problem now, with fossil fuels largely consigned to the dustbin, renewables dominant and emissions under control and on the way down. But this never happened. Instead, despite successive IPCC assessments flagging the increasing urgency of the threat, serious action to tackle what is now, without doubt, an emergency situation failed to materialise.

The fifth IPCC assessment was released in 2014 with the aim of informing negotiations in the run-up to the COP21 summit in Paris the following year. The meeting was hailed an unqualified success – at least by those involved. Amid much cheering and back-slapping, the representatives of 146 countries signed up to the Paris Climate Accord (now ratified by 194 nations), which required them to submit so-called action plans for emissions reductions and promise to 'pursue efforts' to limit the global average temperature rise since Arkwright's time to 1.5°C. These efforts, such as they are, have been ineffective.

In August 2021, the IPCC released the first part of its sixth assessment report, 'Climate Change 2021: The Physical Science Basis' – the remaining three parts are to come in 2022. This initial publication addressed the current state of our climate and what we can expect of it in the decades to come, and its timing was designed to inform discussion and debate at COP26 in late 2021. Because the contents of IPCC assessments are subject to political oversight by representatives of more than 190 nations, the findings presented therein are regarded by some, and rightly so, as being consensus-based and conservative. If anything, this made the conclusions of the latest report that much more frightening, and it is not an exaggeration to say that its publication changed the landscape of the global heating debate for ever.

This time, the IPCC pulled few punches and issued a blunt message. Due to the carbon already released by human activities, it warned, major climate changes are inevitable and irreversible on the scale of a human lifetime. Furthermore, in the absence of immediate, rapid and large-scale reductions in emissions, limiting the global average temperature rise, since Arkwright's time, to 1.5°C or even 2°C would be impossible.

Glasgow and beyond

For all the impact it had on negotiations at COP26, you might wonder if those involved in the summit had even seen the IPCC sixth report, let alone digested its terrifying message. Post-Glasgow, therefore, keeping this side of the 1.5°C dangerous climate breakdown guardrail remains practically impossible. Instead, according to the highly respected Climate Action Tracker, the world continues to follow a path, based upon current policies and action, towards a hothouse future that would see a global average temperature rise of 2.7°C by 2100. This is a 'best' estimate, so the reality could be a bit cooler, or almost a full degree hotter. If the short-term pledges made in Glasgow were made good, the best-estimate temperature rise would still be far too high at 2.4°C, the worst-case reaching 3°C. Even if longer-term goals promised in Glasgow were also met, the rise would still be in excess of 2°C. While these hikes may not seem large, remember that they are global averages that – as I will discuss later – disguise the true picture. Be in no doubt, anything above 1.5°C will see the advent of a world plagued by intense summer heat, extreme drought, devastating floods, reduced crop yields, rapidly melting ice sheets and surging sea levels. A rise of 2°C and above will seriously threaten the stability of global society.

The depressing news is that, as of April 2022, none of the world's biggest economies – which together generate 80 per cent of carbon emissions – are on target to keep the promises they made in Paris to stop the global average temperature rise topping 1.5°C. The UK, along with Costa Rica, Nepal and a handful of small African nations have made commitments that are almost compatible with achieving this goal, but they aren't there yet. Indeed, the UK is showing worrying signs of backsliding. Meanwhile, countries like Australia, Brazil, Canada, China, Germany and the United States continue to develop fossil fuel infrastructure, making it impossible to achieve the emissions cuts needed to fend off dangerous climate breakdown.

More than 140 countries have so far committed to delivering so-called net zero greenhouse gas emissions, meaning they will cut their carbon output significantly and offset the rest by planting trees or adopting some other means of absorbing excess emissions. Some have posted net zero targets that are so distant as to be almost meaningless, although more than 130 have coalesced around 2050. Even so, this is far too late. The global average temperature rise is slated to exceed 1.5°C within a decade, and there is – according to the UK Met Office – a 10 per cent chance that at least one year could see this temperature exceeded by 2023.

It now seems to be pretty much accepted by national governments that the world will overshoot the 1.5°C guardrail much sooner than even the earliest net zero targets. As a result, there is increasing talk of resorting to technological fixes to suck up any excess carbon still being pumped out by the net zero target year in order to attempt, eventually, to bring the global average temperature rise down to below 1.5°C. The problem is that such fixes don't yet exist at anywhere near the scale required and – as I will discuss later – all those proposed are costly, environmentally damaging, risky or downright dangerous.

Most critically, many countries, despite being supposedly committed to achieving net zero by the middle of the century, have no road map to show how they will get there. Inevitably, this translates into a lack of serious action to cut emissions in the near-term (up to 2030) showing up their longer-term net zero plans to be little more than a lot of (ahem!) hot air.

Nasty surprises

While the world's climate scientists are making a reasonable fist of forecasting how far temperatures will rise and how quickly, getting it right is – due to the complexity of the climate system and the interconnectedness of its different elements – a huge ask. Even the most optimistic would agree that there will be surprises in store as our world continues to heat up, and probably not good ones.

One thing that keeps climate researchers awake at night is the idea that we have passed one or more points of no return, or 'tipping points' as they are known in the trade. Bringing temperatures down on an overheating planet is a bit like trying to turn the *Titanic*, and there may well be situations where – whatever we do in terms of slashing emissions – we can no longer avoid the iceberg.

A good, and especially apposite, example is the Greenland

Ice Sheet, the wholesale melting of which threatens a major hike in global sea level. At more than a quarter of a *trillion* tonnes a year, the current rate of ice loss is already astonishing. But there could be worse news. Some researchers think that we may already have passed, or be close to passing, the ice sheet's tipping point. This would mean that, even if temperatures stopped rising today, melting would continue to accelerate until Greenland was ice-free and sea levels around 7 metres higher as a result. This would not necessarily happen rapidly, but it would be baked-in, and so ultimately inevitable.

Then there are so-called positive feedbacks loops. These involve responses to rising temperatures that act to heat up the planet even more. For example, as temperatures have climbed at high latitudes, so the vast areas of permafrost in Arctic Canada and Siberia have started to thaw. This in turn is releasing methane trapped below, which acts – in a self-sustaining loop – to reinforce warming.

Another example is the progressive disappearance of Arctic sea ice. As our planet continues to heat up, so the area covered by sea ice reduces and is replaced by open ocean. Because dark water absorbs more heat than white ice, the more ice is lost, the more temperatures at high latitudes increase, leading to even more melting.

There are plenty of other feedback loops too, all of which act to augment warming rather than suppress it. The problem is that the ultimate impact of such loops, and the timescales over which they operate, are not fully understood so that their influence on how far global temperatures will rise, and how quickly, is poorly constrained. The lesson to take from this is that quoted forecasts for future temperature rises should be taken as minimum rather than maximum estimates.

EARTH'S CLIMATE SWITCHBACK

2

Goldilocks planet

Our planet has evolved over time to be, like baby bear's porridge, just about perfect, or it was until modern humans arrived on the scene. It orbits a well-behaved star, some 150 million kilometres out, in what is commonly known as the Goldilocks Zone. Conditions here ensure that water is primarily liquid, the climate of our world broadly benign and the environment perfect for the emergence and persistence of life.

Like the other seven planets that accompany our world in its perpetual journey about the Sun, the Earth formed within a cloud of gas and dust that collapsed to form a flattened disc around 4.6 billion years ago. Our planet, and the other large rocky bodies in the solar system, were constructed over a period of tens to hundreds of millions of years, by the sweeping up of debris within this disc, a process known as 'accretion'. At some point during this planet-building activity, it seems as if the proto-Earth collided with a Mars-sized object – sometimes referred to as Theia – which tore our growing world apart to form a debris disc from which our world reconsolidated and the Moon was formed.

By 4.5 billion years ago, following on from this cosmic cataclysm, our planet's surface was completely covered by a magma ocean that cooled slowly to form a solid surface. This early crust took a battering from asteroids left over from the accretion process, and again during a later episode of pounding between 4 and 3.8 billion years ago, known as the Late Heavy Bombardment.

More than 100 impacts by objects in excess of 100 kilometres across probably left the Earth's surface looking more like the craggy, pock-marked lunar landscape than the muted blue and green prospect we are familiar with today. Some of the impacting objects would have been comets, bodies of rock and water ice, which brought with them some of the building blocks of the oceans that now cover more than two-thirds of our world.

The early Earth could not really be said to have had a climate, at least not one we would recognise today. The latest research suggests that the atmosphere was largely oxygen-free and made up mainly of carbon dioxide, with a bit of nitrogen and water thrown in. Atmospheric pressure was also as much as 100 times higher than it is at present, so – all in all – it was not a very nice place.

At the time, the Earth and Venus – our twin, which orbits 38 million kilometres closer to the Sun – were characterised by very similar conditions. Both would have had baking surface temperatures insulated by the strong greenhouse effect arising from their carbon-soaked atmospheres. But over time, the conditions on the two planets diverged dramatically. While the higher temperatures due to its closer proximity to the Sun meant that Venus had its water boiled off, our world – set fair in the Goldilocks Zone – hung on to its water in the form of oceans. These, in turn, absorbed much of the carbon dioxide, subduing the greenhouse effect and hauling down surface temperatures. Venus became a hellhole, a place where it is hot enough to melt lead, but, here on Earth, increasingly favourable conditions brought forth life and, ultimately, ourselves.

It is testimony to the importance of our planet's location, slap bang in the middle of the Goldilocks Zone, that life made an appearance so early in its history. Tiny, tubular forms found in rocks as much as 4.28 billion years old are claimed to be the fossilised remains of organisms that lived around hydrothermal vents in oceans that probably formed not much more than a 100 million years earlier. If true, then life must have sprung up extraordinarily rapidly. Even while our world was still being pounded by giant asteroids and comets, the biosphere was already a thing, its primitive and tenacious early forms clinging to existence in the face of global geological mayhem.

Life may have started early, but it was a very long time before it became commonplace, and it needed an oxygen-rich atmosphere to do so. Over a period of half a billion years or thereabouts, single-celled, photosynthesising cyanobacteria gradually pumped up the level of atmospheric oxygen on Earth – a development known as the Great Oxidation Event. The higher levels of the gas proved deadly to many early life forms, but new ones that depended upon oxygen sprang up to take their place. Multicellular forms first appeared on the scene a couple of billion years ago, and life hasn't looked back since.

Today, after more than 4 billion years of unconscious

experimentation, the different elements of our world – the atmosphere and oceans, the solid Earth beneath our feet and the life that is all around us – have come to an innate arrangement that keeps our planet's environment in balance. The eminent chemist James Lovelock calls this set-up Gaia – after the ancestral mother of all life in Greek mythology – and it operates as a kind of superorganism through a system of self-regulating checks and balances that work together to sustain a habitat favourable to the maintenance of life.

The problem is that Gaia is now sick and getting sicker. While taking ice ages and other natural climate shocks in its stride, widespread environmental damage and diversity loss has meant that Gaia is struggling to handle the vast quantities of carbon being pumped out by humankind's activities at a rate unprecedented in Earth history. Lovelock himself is pessimistic that Gaia can get on top of the situation in the short term, and he has expressed the view that civilisation will be hard-pressed to survive the ongoing breakdown of our climate. It is a view that we would be well advised to take heed of in the critical decades that lie ahead.

All change

The road from planet-wide magma ocean to the climatically comfortable world that gave birth to humankind has been long and bumpy. During this interminable journey, our planet has, for all sorts of reasons, switched from icehouse to hothouse and back again on many occasions.

Broadly speaking, the Earth's climate is modulated by a natural thermostat with just three settings, which correspond to different conditions and temperature regimes.

When set on 'greenhouse', our world is ice-free and tropical conditions extend even to the poles. This has been the case for three-quarters or more of our world's long history. For much of the rest of the time, the thermostat has been set on 'fridge', resulting in lower temperatures and the growth of ice sheets at the poles, sometimes extending down to lower latitudes. The most extreme setting is 'freezer', which sees the Earth becoming entirely or almost entirely covered with ice. This setting has only been clicked on twice, both occasions – fortunately for us – long before we appeared on the scene. The first time was during the Huronian glaciation, between 2.4 and 2.1 billion years ago, when ice may have covered the entire surface of the world, leading to a so-called Snowball Earth state.

The cause is widely held to be a weakening of the greenhouse effect as increasing amounts of atmospheric oxygen, exhaled by cyanobacteria during the Great Oxidation Event, reacted with atmospheric methane, breaking it down into water and carbon dioxide. Although both of these are greenhouse gases, they are nowhere near as good at retaining heat as methane, so the result was a progressive and deep cooling. Falling temperatures would have been helped considerably by the fact that the Sun was much fainter at the time, with its brightness down 16 per cent on today's levels.

The thermostat was turned to freezer setting again between 720 and 635 million years ago, during the appropriately named Cryogenian Period. This was the deepest, coldest ice age our planet has ever experienced, with the global average temperature reaching $-12°C$ and equatorial temperatures as low as in the Antarctic today. Repeated pulses of severe cold resulted in the planet being largely covered by ice on a number of occasions, although recent research suggests that there was still open water in equatorial regions. The drivers of this so-called Slushball Earth are complex and disputed. Certainly, the Sun was still dimmer, by about 6 per cent compared to today, which would have helped cooling. Other than this, the primary cause is thought to be, once again, a big drawdown in carbon dioxide levels, sucked from the atmosphere either by weathering processes – which extracted carbon via chemical reactions between rocks and the atmosphere – or by new, more complex life forms using it to build their skeletons.

Since the beginning of the Cambrian Period, which began 541 million years ago, the Earth's thermostat has been stuck, for most of the time, on the ice-free greenhouse setting. The Cambrian is perhaps best known for the extraordinary diversification of lifeforms, known as the Cambrian Explosion, which is likely to have been at least partly facilitated by the warm, shallow seas that covered much of the planet. The thermostat was briefly flicked to fridge on two occasions during this period, bringing ice-age conditions that persisted relatively briefly, but for most of the time global temperatures averaged 27°C or more, sometimes rising above a steamy 30°C, compared to just 14.9°C today.

Notwithstanding a rare cooler episode, greenhouse conditions reigned from the beginning of the Triassic Period, 250 million years back, to the massive asteroid impact that ended the Cretaceous Period 66 million years ago, a time span that coincided almost exactly with the appearance, rise and demise of the dinosaurs.

Shortly thereafter, at least on a geological timescale, the global average temperature spiked at a time known as the Palaeocene–Eocene Thermal Maximum (PETM) – more on this later – after which it was all downhill. Our world's climate has been broadly following a cooling path for the past 50 million years or so, as the thermostat has inched towards fridge. The first glaciers had formed in Antarctica by around 45 million years ago, and 20 million years on the Antarctic Ice Sheet was well and truly established. As planetary cooling continued, so ice began to conquer the Arctic region, including Greenland, around 7 million years later.

This extended period of cooling is likely to have been caused – to a large degree – by the formation of the Himalayas, as the Indian tectonic plate crashed into Eurasia. Rapid uplift of this colossal mountain range resulted in a huge increase in chemical weathering that removed carbon dioxide from the air, progressively reducing the insulating effect of the atmosphere.

By 2.6 million years ago, our world was in the grip of what we know as the Ice Age, also known as the Pleistocene. During this time, the glaciers expanded beyond their polar fastnesses on numerous occasions – eight in the last 750,000 years alone – bringing frigid conditions to temperate latitudes. But this is not to say that the climate was bitter all the time. In fact, glacial episodes alternated with so-called interglacials, during which the ice retreated, and temperatures were often as warm, if not warmer, than today.

The advance and retreat of the ice was far from random, and the timings can be explained by cyclic variations in the Earth's wavering passage around the Sun. During its journey through space, our planet wobbles in a number of different ways and on a range of timescales, which together result in long-term variations in the geometry of the Earth's orbit and axis of rotation. Not only are these variations predictable, but they fit perfectly with the timing of glacial advances and retreats.

The last glacial episode peaked just 20,000 years ago, following which time our planet has been warming again. Officially, we entered a new interglacial – known as the Holocene (from the Greek for 'completely new') – a little under 12,000 years ago, but that isn't the end of the story. We are still in the Ice Age, and the cold is slated to return – if nature were left to its own devices – probably within 10,000 years. The thing is, global heating, driven by human activities, has put a spanner in the works to the extent that the next glacial episode is virtually certain to be postponed, perhaps indefinitely.

The key message we need to take away from this roller coaster ride through Earth history is that, while other factors may be involved, including variations in solar output, the geometry of the Earth's axial tilt and orbit and the disposition of the continents, it has always been the level of greenhouse gases in the atmosphere – particularly carbon dioxide – that has determined whether our climate is warm or cold. As carbon dioxide concentrations touch 420 ppm, and continue their upward path, it is a message we ignore at our peril.

From Narnia to Eden

At the height of the last glacial episode, the level of carbon dioxide in the atmosphere was a lowly 180 ppm. Just 8,000 years later – the blink of an eye when measured against the great span of geological time – it was 260 ppm.

Twenty thousand years ago, the global average temperature was at least 6°C lower than it is today, ice sheets kilometres thick buried large parts of North and South

America, Europe and Asia, and sea levels were down 130 metres. By a little under 12,000 years ago, the Earth had undergone an almost miraculous transformation, which saw our planet metamorphose into the clement world upon which our civilisation has flourished. One of the most dynamic periods in the history of our world saw rocketing temperatures melt the great continental ice sheets like cheese under a grill, pouring prodigious volumes of meltwater into the ocean basins – 52 million cubic kilometres in all.

The environmental and climatic mayhem that defined the transition from Narnia to Eden marked the very end of the Pleistocene Period. What followed could not have been more different. In sharp contrast, at least until human activities intervened, the Holocene has been marked by a relatively stable climate with little variation over time in either global temperature or the level of atmospheric carbon dioxide.

Even so, there was still plenty going on. The dregs of the great ice sheets continued to melt, bringing the seas up to present-day levels, mostly during the early Holocene. At the same time, those parts of the Earth's crust that had been severely depressed beneath the ice at high latitudes were bouncing back rapidly – a mechanism known as post-glacial rebound. This effect spawned a vigorous response from the solid Earth, with magnitude 8+ earthquakes shaking Scandinavia, and up to a 100-fold rise in volcanic eruption rates in Iceland.

Post-glacial rebound continues today in those regions once covered by the great continental ice sheets. In apparent defiance of the global trend, uplift rates remain high enough in places – for example, along the coasts of Sweden and Finland – to give the appearance that sea level is currently falling. This, however, will not last, as accelerating sea level rise, driven by global heating, will soon overwhelm the rate of bounce back.

It is no coincidence that the extraordinarily rapid rise of humankind occurred during the Holocene, the ameliorating climate encouraging the switch from hunting and gathering to farming. This, in turn, brought people together in larger communities that led ultimately to the establishment of the first cities. While certainly stable in comparison with the climate of the preceding Pleistocene Period, the Holocene has still thrown up surprises, some of which are thought to have played critical roles in spurring on the development of human civilisation.

Most notable was a cold snap known – somewhat uninspiringly – as the 8.2 ka event ('ka' being shorthand for 'kilo-anni' or thousands of years, because it happened that long ago). The cause was the catastrophic emptying of a huge and expanding glacial lake in North America into the North Atlantic. The shock of an influx of some 160,000 cubic kilometres of cold freshwater raised sea levels by up to 4 metres almost overnight and slowed or stopped the Gulf Stream, triggering a centuries-long spell of cooling.

Temperatures fell by as much as 5°C in some parts of the world, bringing to parts of Asia and Africa drought conditions that may have lasted for centuries. While extreme drought is never welcome, in this case it seems – as far as societal development is concerned – to have been a blessing in disguise. This is because it looks as if the paucity of water may have encouraged the development of irrigation in ancient Mesopotamia and the bringing together of people in larger social groupings to cope better with food scarcity. The timing certainly seems to jibe with demographic changes across the region that involve the growth of bigger communities.

The 8.2 ka event was one of a series of cold snaps, known as Bond Events, which intervened periodically to reduce the general warmth of the Holocene. The underlying cause of the others is not certain but may be related to cyclical variations in the North Atlantic currents, or to episodic reductions in the Sun's activity. Certainly, the most recent event, a modest cooling of the northern hemisphere between the 16th and 19th centuries known as the Little Ice Age, does broadly coincide with a period of reduced solar output known as the Maunder Minimum.

Taking the Holocene as a whole, the global average temperature followed a rising trend until around 5,000 years ago. Then, during the so-called Holocene Climatic Optimum, summer temperatures in the northern hemisphere may have been as warm as they are today. From then on, temperatures began to fall and continued to drop until the start of the 20th century when human-driven global heating got going in earnest, cancelling out the natural cooling of the previous five millennia.

Today, instead of looking out for the arrival of a new glacial period, we look around us at a world of blistering heatwaves, raging wildfires and unprecedented droughts. You would be forgiven for thinking that we must be coming to the end of the clement conditions of the Holocene and entering another period of climate instability and chaos, and you would be right. In fact, many researchers, from across the whole spectrum of the sciences, have been arguing for some time that we are now in a new geological period. They have even come up with a name, the Anthropocene – after the Greek 'anthropos', for 'human', and 'cene', meaning 'recent' or 'new'. This name is completely appropriate for a time when humankind's signature is written across every corner of the natural world; a time when our pollution, be it carbon dioxide, radioactive isotopes or microplastics, infiltrates and contaminates everything, down to pristine Antarctic ice and the placentas of pregnant women.

For the moment, the Anthropocene remains a concept without official recognition, but it is only a matter of time before it is given the nod as the name for the next geological period after the Holocene.

Debate and discussion abounds, among advocates for this new period, about when its start date should be, given that – as we shall see in the next section – humans have been making their mark on the environment, intentionally or otherwise, for thousands of years. Most, however, agree that the mid-20th century should mark the beginning of the Anthropocene – its onset delineated in the geological record by the occurrence, in new sediments, of radionuclides from atomic blasts and reactor accidents, heavy metals from industrial processes and ubiquitous microplastics. Whatever date is finally agreed upon, there can be no doubt that humankind is now the planet's dominant force, moulding and bending nature to its will.

The human greenhouse

As I noted briefly earlier, there is nothing inherently wrong with greenhouse gases – indeed, they are vital constituents of our world's atmosphere. Working together, carbon dioxide, methane, nitrous oxides, water vapour and the other greenhouse gases result in the greenhouse effect, which acts to insulate our world from the bitter cold of space. Without it, our planet would be a lifeless iceball.

Greenhouse gases act, then, in a similar way to a common-or-garden greenhouse or conservatory. They let incoming solar radiation pass through and reach the Earth's surface, but they stop heat travelling back out into space. As tomatoes and peppers benefit from the warmth that builds within a greenhouse on a sunny day, so life on Earth has benefited from the raised temperatures caused by atmospheric carbon dioxide and the other greenhouse gases. The problem is that in pushing up the concentrations of these gases, we are increasing surface temperatures on Earth to a degree that is causing our stable climate to break down. What we are doing is a bit like building a second greenhouse – a human greenhouse – around the natural one.

Earlier, I placed the blame for lighting the touchpaper of human-induced global heating squarely at the feet of Richard Arkwright. This was probably a little unfair. The truth is that humans have been affecting the composition of the atmosphere for many thousands of years.

Between 10,000 and 15,000 years ago, the numbers of mammoths and mastodons – nature's kings of the Ice Age – plummeted. Undoubtedly, the rapid post-glacial warming had a great deal to do with this. It has been argued, however, that groups of well-organised human hunters armed with spears also played a key role. As the mammoths and their relations were consigned to evolution's dustbin, so – it seems – the scrubby vegetation, bushes and small trees that they munched upon and kept in check began to flourish and spread, quickly replacing the grasslands of the Canadian and Russian tundra. Because darker scrub cover absorbs more solar radiation than paler coloured grasses, the temperature of these regions may have climbed by 0.2°C and as much as 1°C in places. If, then, humans did play a role in the demise of these majestic animals, it means that this was the first occasion on which our race meddled in a wholesale manner, albeit unintentionally, with its environment, marking the very start of global heating due to human activities.

Jump forward a few thousand years and the hunter-gatherers of the time began to lose interest in chasing mammoths, presumably because there weren't that many left by this stage. Encouraged by the warm and stable growing conditions, and seeking a more reliable food supply, they began to settle down in groups to husband animals and farm crops. In the Middle East's 'fertile crescent', between the waters of the Euphrates and Tigris, this began around 10,500 years ago, but the same idea caught on independently in many other parts of the world.

By around 8,000 years ago, farming was becoming widespread, and to make space our distant ancestors were hacking down trees left, right and centre, resulting in massive deforestation across Europe and Asia. With wood burning carried out on an industrial scale and fewer trees to soak up carbon, the levels of carbon dioxide began to rise, from around 260 ppm to 280 ppm, the level at which they stayed until Arkwright came along.

Put the clock forward another few thousand years and atmospheric methane levels began to climb too. This time, the increasingly widespread cultivation of rice, particularly across Asia, has been fingered as the culprit. The soggy soils beneath flooded paddy fields are especially good at producing this potent greenhouse gas, and rice production today is responsible for around 10 per cent of agricultural methane emissions.

Altogether, it is estimated that the prehistoric switch from a hunter-gatherer lifestyle to one of relatively sedentary farming resulted in a global average temperature hike of a shade under 1°C, and perhaps up to 2°C at high latitudes. Considering that the entire population of the planet at the time may have been just 5 million or so, this is an extraordinary feat. It also throws into stark relief the potential climate mayhem we can wreak with an early 21st-century population just shy of the 8-billion mark.

If the pattern of carbon dioxide levels during other interglacials is anything to go by, the amount of the gas in our atmosphere should have been falling for the past 10,000 years or so and should, by now, be below 250 ppm. Instead, of course, it is going through the roof. As noted previously, carbon dioxide levels in the atmosphere peaked at a little under 420 ppm in 2021, up 50 per cent on Arkwright's time, and the highest concentration for around 15 million years. Unless drastic action is taken to reduce emissions, we can expect the level of carbon dioxide in the atmosphere to reach 560 ppm – twice the pre-industrial concentration – by the 2070s, or even earlier, bringing the prospect of severe and sustained hothouse conditions on our planet.

So long as we persist in growing the insulating shroud of greenhouse gases around the Earth, so the rate of melting of the polar ice sheets will continue to accelerate, sea levels will continue to increase and extreme weather will exact an ever greater toll. The longer we maintain the human greenhouse, the more perilous the future of our race will become, the more drastic will be the measures required to bring greenhouse gas levels down and the longer it will take to rewild our climate back to its natural interglacial state.

HOT AND STEAMY WITH A CHANCE OF COLLAPSING ICE SHEETS 3

Earth's climate today

On 29 June 2021, the unassuming Canadian village of Lytton, in southern British Columbia, registered an astonishing temperature of 49.6°C (121°F), beating the previous record for the highest temperature ever recorded in the country by the huge margin of almost 5°C. It was also the highest temperature ever recorded north of the 50th parallel, and hotter than anything ever experienced in Europe or South America. The following day, the village was gone, wiped from the face of the Earth by one of the many wildfires triggered by the searing temperatures.

Barely two weeks later, on 12 July, a slow-moving line of thunderstorms dumped up to a month's worth of rain on London, bringing widespread flash flooding that caused sewers to back up and travel chaos. But far worse was to follow. Over the next three days, the same low-pressure system stalled over eastern Belgium, Luxembourg and western Germany. Warm, moist air sucked up from the south fed a biblical deluge that brought the worst floods in 1,000 years to some parts of the region and unprecedented and shocking scenes as the power of water devastated well-off communities in the heart of Europe.

Without warning, the wild weather cultivated by global heating wasn't battering some distant land but taking its awful toll just across the Channel. Suddenly, it was too close for comfort; all too easy to imagine the images of raging torrents, stranded corpses and demolished homes transposed to our own communities.

More than 1,000 lives were lost as a direct consequence of the unprecedented North American heatwave, while the European floods took close to 250 lives and caused damage totalling in excess of $11 billion. But these instances were just two in a long line of extreme weather events in 2021 that destroyed lives and livelihoods right across the planet. Devastating floods also swamped great tracts of Turkey, China, Japan, India, Pakistan, the United States and New Zealand. Meanwhile, some of the greatest wildfires ever seen raged across Siberia and California, while record-breaking droughts became even further entrenched across the western United States, Central Asia and southern Africa. It was not only Canada that experienced unparalleled heat. All-time temperature records were shattered across much of North America and Southern Europe. Sicily smashed the European record with 48.8°C (120°F), while the heat in California's appropriately named Death Valley touched 54.4°C (130°F), the highest temperature ever (reliably) measured on the planet.

The reality is, then, that we don't need to look beyond our news feeds and TV screens to build a picture of climate breakdown today. While a relatively small rise in the global average temperature can only be detected using instrumentation, its ramifications, in the form of extreme weather events, are now out there for all to see.

I can hear the more sceptical among you pointing out that there have always been bouts of severe weather and always will be. This is true, but the fact that weather records are being broken left, right and centre – at a record rate in fact – tells us that something unusual is at work here. Furthermore, it is now possible to calculate how likely a particular weather event would have been with and without global heating. The 2021 European floods, for example, were nine times more likely to occur than in a world where global heating had never happened. Even more clear-cut, the extraordinary 'heat dome' that roasted much of western North America during the early summer of 2021 would have been virtually impossible in a world where human activities had not artificially pumped up atmospheric carbon levels. In fact, this unprecedented heatwave was a one in 1,000-year event, made 150 times more likely by global heating. It is a disturbing thought that fully 70 per cent of more than 400 extreme weather events, analysed by researchers, were found to have been made more likely or more severe as a result of global heating.

Perhaps the most worrying thing about the obvious boost in the frequency and intensity of unprecedented weather events is that this is being driven by a relatively small temperature rise. The past eight years (to 2022) have been the hottest ever recorded, while the global average temperature over the past twenty years exceeded 1°C more than in Arkwright's time. Evidence that global heating doesn't stand still is provided by the fact that the temperature rise in 2021 was a full 1.2°C above pre-industrial levels.

But the hike in global average temperature is far from the whole story. No part of the Earth environment has been shielded from the effects of global heating, and the assessment reports released by the IPCC check off the manifold consequences of the modest rise in temperature that have become apparent so far.

It is no surprise that heatwaves have become more intense and are lasting longer. As a hotter atmosphere contains more water in the form of vapour, rainfall across the land has increased as our world has heated up, and more rain is falling as intense downpours so that the incidence of serious flooding is rising too. Storm tracks have migrated towards higher latitudes, while heating of the oceans has increased their potential to spawn and drive more powerful hurricanes. Hotter seas are also taking their toll on coral reefs, and on land many species are struggling to keep up with the relentless poleward march of climate and vegetation zones at up to 10 kilometres a year, as temperatures at a particular latitude rise ever higher. This is leading to enforced migration and dietary changes, breeding issues, problems with pollination and – in many cases – diminishing numbers. As would be expected, glaciers across the world have gone into rapid retreat and the area covered by Arctic sea ice has diminished by up to 40 per cent since the 1980s. The rate of sea level rise is now 5 millimetres a year, reflecting accelerating melting of the Greenland and Antarctic Ice Sheets and the tendency of the oceans to take up more space due to thermal expansion. Half a centimetre may sound like peanuts, but it is getting on for four times the 1900–1990 rate and continuing rapid acceleration could see a far higher figure within just a few decades.

Taken together, all this tells us that climate breakdown is not something that belongs to the distant future. On the contrary, it is with us now, today. It is no exaggeration to say that the speedy burgeoning of severe weather reveals that our climate is on the very cusp of radical change that will transform our planet so that extreme, hothouse conditions are commonplace. Just how extreme is entirely within our gift to determine.

The devil in the detail

Halfway between the Norwegian mainland and the North Pole is the frigid island group known as Svalbard (once better known as Spitsbergen). Here, the Sun sets on 26 October and doesn't make an appearance again until 15 February and signs warn drivers to beware of polar bears on the road.

Svalbard also hosts Longyearbyen, the world's most northernmost town, which also happens to take the prize for the fastest heating place on the planet. In fact, since 1971, temperatures here have climbed by an extraordinary 4°C, which is *five times* faster than the global average. If it keeps going at this rate, the town, along with the entire archipelago, is predicted to be an eye-watering 10°C hotter by the century's end.

The people of Longyearbyen are already suffering the consequences of global heating. The climate is getting wetter, with rain often falling instead of snow, and wilder, as storms replace the icy calm of the bitter winter night. Thawing permafrost is damaging buildings and husky-drawn sledges are as likely to have to use wheels as runners. In winter it used to be possible – in theory at least – to walk to the North Pole across the pack ice, but no longer. The glaciers, which cover around 60 per cent of Svalbard, are retreating rapidly, raising the risk of flooding. Avalanche activity is increasing too, as rain and wet snow make slopes unstable. In 2015, an avalanche from Longyearbyen's local mountain, Sukkertoppen (Sugar Top), crashed into the town's suburbs at night, killing two and injuring eight sleeping inhabitants.

What is happening in Svalbard signals an important aspect of global heating. Not everywhere is heating at the same rate. This is disguised by the common usage – in discussions about climate breakdown – of the global *average* temperature rise. So, even now, at the height of the climate emergency, some people, when told that our world has already heated up by more than 1°C, seem quite unconcerned. What's 1°C among friends is their attitude. Surely it is not enough to worry about and, in any case, maybe a little bit more heat would be a welcome antidote – say some Brits – to our often-soggy summers.

The situation on Svalbard teaches us that the seemingly innocuous global average figure is one that misleads, that encourages complacency and that hides much of the detail about what is really happening. As climate scientists are fond of saying, no one lives at the global average temperature. The reality is that different parts of the planet are heating up at very different rates. Some, like parts of the south-east United States, have followed cooling trends, although this will not continue in the long term.

Luckily for us, the oceans have done a sterling job of soaking up more than 90 per cent of the heat generated by human activities in the past half century – at an extraordinary rate equivalent to devouring the heat from 440 billion toasters, 24 hours a day, 365 days a year. Even so, due to the colossal volume of the oceans and the fact that it takes much more energy to heat water than air, the sea surface temperature is only a little less than 0.9°C hotter now than it was over the baseline period of 1850–1900, although the rate of increase is now accelerating.

On the other hand – and far more relevant because this is where we all live – temperatures over land have climbed far more rapidly over the same period, by 1.6°C on average. Due to the way heat is distributed by marine currents, and because a much greater area is land, the northern hemisphere has heated up more quickly than the southern hemisphere, which is mostly ocean. The higher figure means that the impact of global heating is significantly magnified across northern hemisphere land masses, most notably via more frequent, more intense and longer lasting heatwaves and droughts. In the northern hemisphere, the loss of ice is also aggravated so that mountain glaciers are in rapid retreat and the permafrost that holds rock faces together is thawing, leading to increasing rock fall and landslide risk.

And even the world's land masses are not heating equally rapidly. Those at high latitudes, such as Svalbard, are experiencing far higher temperature rises than those in the tropics, due to what is known as polar amplification of heating. There are a number of possible causes, but a key one is the replacement of white ice by darker open sea or land, which results in more of the Sun's heat being absorbed. Greater heating at high latitudes is hugely significant because this is where most of the planet's ice and frozen ground reside. Consequently, higher temperatures here bring the prospect of ice sheet collapse, big sea level hikes and permafrost methane outbursts that have the potential to accelerate global heating even more.

Like Svalbard, the Antarctic Peninsula – the long finger of land that points northwards towards Chile – has been heating *five times* faster than the global average, with the temperature shooting up by close to 3°C between 1950 and 2000 alone. At the top of the planet, Alaska has experienced a temperature rise of more than 2°C in the past 50 years, and the rate of heating is accelerating. This is bringing all sorts of problems that would be familiar to the inhabitants of Svalbard, including building and road subsidence as the permafrost thaws and an increasing threat of landslides as mountain ice melts.

In 2019, across Alaska, 32 high temperature records were smashed, with Anchorage seeing a high of 32°C (90°F) for the first time ever. In 2021, temperatures exceeded 27°C (81°F) at several locations and touched 31°C (88°F) in Fairbanks – another all-time record. It is hard to believe, I know, but wildfires are becoming an increasing problem across the state better known for its snow and ice.

The devil, then, is very much in the detail. And the detail reveals that – as far as future prospects are concerned – our world is heating up in the worst possible way. Those regions with the most ice are experiencing the biggest temperature rises, so maximising the potential for wholesale melting and large, rapid hikes in sea level. As land temperatures climb faster than those of the ocean, our lives and livelihoods bear the brunt, while the shock to heat-sensitive environments, such as rainforests and the Arctic tundra, is maximised. All in all, it is not a pretty picture.

Broken climate

There can be no doubt that our once stable climate is broken, and evidence – in the form of escalating extreme weather – is now all around us. But how are the two related? Just how does global heating drive changes to our weather on a seasonal, monthly and even daily basis? There is no single answer, but in broad terms, a failing climate translates into extreme weather via big changes to global weather patterns. This can encourage the development of persistent, extreme heat, but it can also bring severe cold.

Drilling down further, a good place to start is in the Arctic, where global heating is hitting hardest. It will come as no surprise to hear that the air above the North Pole is very cold. This frigid air is contained and isolated from the warmer atmosphere further south by a narrow band of fast-moving high-altitude winds known as the jet stream, which flows from west to east at heights of between 8 and 12 kilometres.

The North Polar Jet Stream (there is one in the southern hemisphere too) holds in the cold air and can travel at speeds in excess of 400 km/h. It often follows a reasonably straight path, but sometimes – like a river – it meanders back and forth. The jet stream owes its strength to the difference in temperature between the cold polar air and the milder air at lower latitudes. The problem is that, due to the melting of sea ice exposing darker ocean that absorbs more heat, the cold Arctic air is getting warmer, so that the temperature gradient between the pole and temperate latitudes is being progressively reduced. The result is a North Polar Jet Stream that seems to be weakening. This, in turn, leads to the stream having a more wildly meandering course capable of bringing biting cold further south and allowing milder air to surge far to the north. Due to accelerating global heating, then, what happens in the Arctic no longer stays in the Arctic.

Sometimes, the kinks in a vigorously meandering jet stream combine to form the shape of the Greek letter omega

(Ω) so that an area of high pressure in the centre is flanked by two areas of low pressure. Such omega blocks are especially effective at halting the usual west to east movement of the weather and – once established – are difficult to shift. Consequently, they can lead to the development of persistent weather conditions that can last for weeks, even months, at a time.

Long-lasting, deep cold, such as that which established itself across much of the United States and Europe in the spring of 2013, is brought about by what are known as polar vortexes – low-pressure systems carrying icy air south from the Arctic around one or other side of an omega block.

On the other hand, persistent hot weather builds beneath the high pressure within the block itself. Examples include the 2003 European heatwave, which took more than 70,000 lives, and the long-lasting heat dome that afflicted much of western North America in the early summer of 2021. In the UK, the 2020 'lockdown' spring – the sunniest on record – was also the result of an omega block that kept high pressure over, or close to, the country from March to May.

Blocking linked to a strongly meandering jet stream can also influence storm tracks, pushing them into unusual trajectories, sometimes with catastrophic results. Hurricane Sandy, for example, which battered New York and New Jersey in October 2012, was guided onshore by a blocking high-pressure system out in the North Atlantic. A similar weather pattern diverted Hurricane Florence towards the state of North Carolina in 2018, leading to widespread coastal and inland flooding.

To make things worse, hurricanes also seem to be slowing down, meaning that their impact on a particular area lasts longer and, therefore, has the potential to cause more damage. Hurricane Harvey, which lingered over Houston and south-east Texas in August 2017, is a classic example, bringing unprecedented rainfall and driving floods that killed more than 100 people. Looking ahead, it is forecast that changes to global wind patterns will result in hurricanes slowing by 10–20 per cent in a hotter world.

The big worry is that continued rapid heating of the Arctic will lead to further shallowing of the temperature gradient between the poles and temperate latitudes, causing more blocking that will, in turn, drive more extreme weather. There are climate models that suggest a hotter world should result in less blocking weather patterns. Contemporary observations do not, however, support this assessment, leading to some climate scientists suggesting that these models are missing something.

Global heating is also driving changes to other climate phenomena that can translate into a further burgeoning of extreme weather. Probably the most notable is the infamous El Niño, a periodic warming of the central and eastern tropical Pacific Ocean, which is our planet's most influential climate phenomenon after the seasons. El Niños develop every two to seven years and normally last for anything from nine to 24 months. They have been around for at least thousands of years, bringing about dramatic changes to global weather patterns that see some places deluged by heavy rains, while others are afflicted by severe drought.

The jury is still out on whether global heating will bring about an increase in the frequency or strength of future El Niños, although one piece of research suggests that the number of extreme 'super' El Niños could double by the end of the century. Notwithstanding this, El Niño occurrences help to enhance the rate at which the planet is heating up, at least temporarily. In 2015, for example, the sixth hottest year on record, a powerful El Niño added 10 per cent to the temperature hike caused by carbon emissions.

Superimposed upon the steady ramping up of temperatures due to global heating, the effects of El Niños are also becoming ever more amplified, so that the impact of extreme weather conditions associated with these events are exaggerated. In this way, the 2015 El Niño magnified prevailing high temperatures and drought conditions, resulting in a reduction in food production that pushed nearly 6 million children across the world into severe hunger.

The corollary of all this is that global heating does not so much intensify severe weather directly as bend and break long-established weather patterns, which translates into new conditions that favour extreme weather events. It may well be that, in time, savage and destructive weather becomes so commonplace as to represent the normal state of affairs. We need to face the fact that, ultimately, there might be no such thing as extreme weather any longer, just weather.

Towards 2100 and beyond

Forecasting how much and how quickly our planet will heat up in the decades and centuries ahead, and how the climate will respond, is staggeringly difficult. This is because it depends not only on the rate at which greenhouse gas emissions fall, if indeed they do fall, but also on how exactly the Earth's natural systems respond. This is especially the case where key tipping points and positive feedback loops are concerned, which remain very poorly understood. When, if at all, for example, will the Greenland and West Antarctic Ice Sheets begin to collapse in earnest? Will the Gulf Stream shut down, bringing plunging temperatures to the North Atlantic region? If so, when will this happen? How much methane will be broadcast into the atmosphere as the Siberian permafrost thaws? Will the oceans ever stop sucking up carbon? These are questions that we can't answer, at least just yet. But they have huge implications for what our world will look like as the century progresses.

When it comes to building a picture of what sort of world our grown-up children and their children will inhabit in later life, the IPCC assessment reports are still regarded as *the* benchmark, although, as I will explain later, they certainly don't have all the answers – far from it. In its sixth assessment of where we are and where we are going, the IPCC considered the response of the climate to five different scenarios, known as Shared Socio-economic Pathways (SSPs), which are based upon the level of greenhouse gas emissions, changes in land use and atmospheric pollution. These scenarios, in turn, plug into modelled projections of how climate breakdown will develop in the short term (2021–2040), medium term (2041–2060) and long term (2061–2100).

There simply isn't the space here to discuss all five models in detail. Suffice to say that they range from the almost naïvely optimistic very low emissions scenario 'SSP1-1.9' to an effectively worse case – some might say the most realistic – very high emissions scenario 'SSP5-8.5'. What all five have in common is that the expected temperature rise by 2040 (relative to 1850–1900) is pretty much the same, ranging from a best estimate of 1.5°C for the most optimistic scenario to just 1.6°C for the worst case. Thereafter, however, the projections diverge dramatically.

For the very low emissions scenario, the best estimate global average temperature rise actually begins to reduce in the last decades of the century. It would be marvellous if this proved to be the case, but we are nowhere near getting on to this emissions trajectory. Furthermore, it does not fit with previously mentioned research that suggests a more than 2°C rise is already unavoidable, whatever we do. For the same reasons, the low emissions scenario, which sees the global average temperature rise stay below 1.8°C this century, is also very unlikely to match the reality.

For the intermediate, high and very high emissions scenarios, the 2°C mark is either certain or extremely likely to be surpassed, and this could come within twenty years. For the most pessimistic scenarios, the global average temperature continues to climb until, by 2081–2100, it is 3.6°C to 4.4°C higher than the 1850–1900 average, perhaps even more.

Broadly speaking, the IPCC figures fit reasonably well with the post-COP26 Climate Action Tracker predictions discussed in the first chapter. Both paint the same dismal picture of a world in which the global average temperature rise will, barring a miracle, climb above 1.5°C and – almost certainly – above the 2°C mark as well. Furthermore, without serious emissions cuts in the immediate future, a world where the global average temperature rise is at least twice that of the 1.5°C dangerous climate breakdown guardrail beckons later this century.

It goes without saying that should the worst-case projections come to pass, we would be in deep, deep trouble. But even if we manage to bring emissions under control, the speed with which we do so is absolutely critical. This is because every degree of temperature rise, even every tenth of a degree, chips away that little bit more at our previously benign climate. A one in 50-year heatwave, for example, would be 8.6 times more likely with a global average temperature rise of 1.5°C, but this increases to fourteen times with an average temperature rise of 2°C. Similarly, heavy rainfall events will become a little over 10.5 per cent wetter in a 1.5°C hotter world, but 14 per cent wetter if the global average temperature rise touches 2°C.

The lesson is, then, the quicker we act to cut emissions, the less violent and destructive will be the whirlwind we and our descendants reap. In fact, the amount of carbon we have emitted up until today is directly linked, in a linear manner, to the rate of increase in global average temperature. And this will apply in the years ahead too, so that every single tonne of carbon released counts equally towards making things worse, and every tonne less will make a difference.

Of course, it is not just the temperature value that we should be worried about. As the century progresses and the planet continues to heat up, so the oceans will become ever more acidic as they soak up more carbon dioxide, threatening coral reefs and other marine organisms, life on land will struggle as weather patterns and climate zones change and the world's ice will keep right on melting.

I like to think of the boundary between sea and land as climate breakdown's front line, because here our success in the battle against global heating can be measured directly in terms of how rapidly the sea is encroaching upon our coastlines. The rate at which the seas are rising is partly a function of thermal expansion as warmer waters take up more space, but it is essentially a proxy measure for how much and how quickly the great polar ice sheets are melting.

In its sixth assessment, the IPCC forecast that, for all scenarios, sea level is unlikely to rise above 1 metre by 2100.

This is a view that has been challenged by some researchers, who have modelled far larger rises. The IPCC does nod in the direction of this view by flagging an additional 'low likelihood – high impact storyline' that, it concedes, could see at least another 0.5-metre rise by the century's end.

Even a 1-metre rise would, however, have a colossal impact on coastal communities. Around 250 million people are estimated to live within 1 metre of the high-tide mark and could be driven out of their homes by a combination of a 1-metre sea level hike alongside more frequent storm surges and tidal floods. Should the sea level rise reach 2 metres by 2100, which is by no means impossible, one estimate predicts that the number of people displaced as a result would be getting on for 500 million in all.

The world of 2100, a world that some of our children and many of our grandchildren will live to see, will be very different from ours. Just how different will depend enormously upon the efforts of world leaders over the coming decade. Inevitably, it will be a hotter world, with higher seas, more extreme weather and the major displacement of populations. Without serious action to slash emissions in the next few years it will be far, far worse.

HOTHOUSE PLANET 4

Back to the future

Modelling is a tried and tested means for predicting our future climate, but the output does depend upon what goes in, so that the adage 'garbage in, garbage out' applies. Of course, climate scientists do their damnedest to ensure that the data that goes into their models is the best available. Nonetheless, there are still so many unknowns associated with global heating that the output can only provide a guide to what our world will look like as we go forward.

There is, however, another way to gauge how global heating might progress, which is to look back to a time when temperatures and/or greenhouse gas levels were similarly elevated and use conditions then as a bellwether to forecast what we might expect to happen in our future. There have been many other heat spikes in our planet's long history, promoted by natural processes, and it is helpful to compare some of these with today's human-driven hike.

The global average temperature between 2011 and 2020

was higher than at any time since before the last ice age – a period 125,000 years ago known as the Eemian interglacial, when temperatures were comparable to today's. This was a time when hippos roamed the Thames Valley and sea levels were at least 6 metres higher than they are today, but the fact that carbon dioxide levels were not as high as they are presently means that this is not an exact analogue for our world.

We need to go back further, to a time around 15 million years ago known as the Middle Miocene Climate Optimum (MMCO) to find carbon dioxide levels comparable to what they are now, perhaps somewhat lower. The alarming news is that the global average temperature back then was between 2–4°C up on Arkwright's time, providing a clue as to where our overheating world is headed, and this fits well with modelled forecasts of global temperatures in the decades to come. Even scarier, sea levels during the MMCO were a shocking 20 metres higher than today, suggesting that whatever action we take to cut emissions, this amount of sea level rise may already be baked-in.

Such comparisons with prehistoric warmth certainly don't inspire any confidence that we can dodge a hothouse future. The real concern, however, lies in the fact that we are not yet done. The rate of emissions still shows no sign of reducing and promises made by the world's governments don't look like translating to a fall, or even a slowdown, any time soon. Analysis of how quickly emissions have risen since Arkwright's day reveals that we have to go back 56 million years to find the last time that greenhouse gases were being released at anything like the current rate. Then, during the PETM, natural processes resulted in a carbon outburst that pushed global temperatures rapidly upwards by around 6–8°C, leading to marine extinctions and major changes to environments and ecosystems. Making the most of the accelerated warmth, crocodiles thrived north of the Arctic Circle, while palm trees adorned parts of Siberia.

The PETM was a time of extraordinary climatic change, which happened – at least in a geological context – astonishingly quickly. Perhaps over a period as short as 20,000 to 50,000 years, more than 44,000 billion tonnes of carbon dioxide is estimated to have been released into the atmosphere. The source is not clear, and continues to be much debated, but two candidates include large-scale volcanic activity or the release of prodigious volumes of methane from beneath the seabed.

What really should make us sit up and take notice is that the average annual rate of carbon dioxide release during the PETM could have been as low as 1 billion tonnes a year, even a little less. Today, human activities are pumping out the gas at a rate around 40 times faster. If this continues, we will match the total amount of carbon dioxide released during the PETM in around 1,000 years. It is perfectly possible that carbon has never been released naturally at such a speed in the whole of Earth history. This puts us well and truly in uncharted territory and marks out our time as quite possibly unique.

And there is more grim news. The PETM lasted for around 200,000 years, an exceedingly long time, the cause of which has puzzled researchers. Now, it seems, the answer may be forthcoming. As the world heated rapidly, severe storms, heavier rainfall and floods became more common, just as we are seeing today. This resulted in rapidly increased weathering and erosion under warm and wet conditions, which – instead of helping to absorb carbon, as happened at the time of Slushball Earth mentioned earlier – actually released huge quantities of carbon into the atmosphere.

Long after the initial burst of carbon that led to the PETM temperature spike, it was this additional long-term carbon release that led to temperatures staying high for hundreds of millennia.

The lesson for us is that in the hotter, wetter world towards which we are headed, the release of carbon into the atmosphere by natural processes may well continue, even if we cut back our own emissions. This has the potential to lock our world into hothouse conditions for many tens of thousands of years. A sobering thought indeed.

We can build models that seek to shed light on the world of the late 21st century and beyond, and we can look back at past climates for clues as to where our world is headed. When all is said and done, however, we must recognise that the current situation is close to unprecedented in our planet's unimaginably long existence. As such, in relation to forecasting what sort of world our children and grand-children will likely inhabit, your guess is as good as mine. Nonetheless, one thing we can be sure of is that summers will be much hotter, and not in a nice way.

Summers that suck

I know there are those of you who will argue otherwise, but there *is* such a thing as too much sun. Sunlight is vital for photosynthesis, for promoting well-being and – these days – for powering solar panels. But as the insulating blanket of greenhouse gases that shrouds our planet thickens, the trapped heat from sunlight is bringing ever-increasing problems. Across the spectrum, from health to agriculture, from water supplies and the stability of mountain glaciers and rock faces to raging wildfires, too much sun makes for a hotter world and is a recipe for disaster.

Since the middle of the 20th century, the northern hemisphere summer has increased in length from 78 days to more than 95. By the end of this century, summers here are forecast to last half a year, with the duration of winter slashed to just eight weeks.

Some of you may remember the blistering heat in the UK summer of 2003. It was one of the reasons I moved north from the capital and up into the cooler uplands of the Derbyshire Dales. It was also one of the first episodes of unprecedented weather that made people sit up and take notice of what global heating was all about. Across Europe as a whole, the intense heat throughout much of July and August led to summer 2003 being the hottest since at least 1500, resulting in the loss of an estimated 70,000 lives. France suffered most, with temperatures in some places topping 40°C for more than a week. In the UK, the all-time heat record was broken in August when the mercury touched 38.5°C in Faversham, Kent. In the Portuguese town of Amareleja, temperatures climbed to 47.4°C.

The 2003 heatwave was the focus of the first ever 'attribution' study, which ascertained the extent to which humans influenced the event. The conclusion was that global heating driven by human activities made the heatwave at least twice as likely as otherwise. A second study found that extreme heatwaves in Europe are now over ten times more likely due to global heating. According to the UK Met Office, such heatwaves could happen every other year by the middle of the century.

Since 2003, devastating heatwaves have become more frequent and more extreme. Zeroing in on a few, the summer of 2010 saw unprecedented, widespread heat across much of Russia, North America, Eastern Europe, the Middle East and China driven by the highest April to June northern hemisphere land temperatures ever recorded. In 2013, Australia's so-called Angry Summer saw more than 120 weather records smashed in just three months, with temperatures threatening the 50°C mark in a number of places. Three years on, pre-monsoon late Spring heat across India in 2016 pushed temperatures to 51°C. And the following year was even worse, with summer temperatures in Kuwait and Iraq registering 54°C and 53.9°C respectively. These remain the highest ever temperatures reliably recorded anywhere on the planet, outside of California's notorious Death Valley. Just four years later, the sweltering summer of 2020 helped make the year the hottest ever recorded in Europe, with the temperature for the year a whopping 1.9°C above the long-term average, an astonishing – in meteorological terms – 0.5°C increase on the previous high. Such a big hike is especially worrying as it flags the possibility that the heat may continue to build rapidly in sudden jumps, rather than climb slowly and incrementally.

The ramping up of hothouse conditions culminated in the searing heat of spring and summer 2021. Temperatures began to climb towards the end of May, reaching close to 32°C within the Arctic Circle and exceeding 35°C across much of Siberia. By mid-June, it was western North America's turn to bear the brunt, with temperatures beneath the unprecedented heat dome reaching 47°C in Portland (Oregon), a shade under 50°C in Lytton (British Columbia) and 54.4°C in Death Valley. As a consequence of this astonishing weather event, which – studies suggest – was made 150 times more likely by global heating, California, Idaho, Nevada, Oregon and Utah recorded their hottest ever summers, while sixteen other US states recorded a top-five hottest summer. As previously touched upon, Europe had its hottest summer ever in 2021 too. Here, the level of heat recorded would normally be expected every 10,000 years, but, astonishingly, the continent is now experiencing such conditions – on average – every three years. Without massive emissions reductions, such sweltering summers will happen every year by 2100.

During the course of 2021, countries hosting a quarter of the world's population experienced their highest temperature on record, but this is far from the end of the story. As the global average temperature continues to climb, so heatwaves will become more frequent, get hotter and last longer. Heatwaves topped the list of global disasters for both 2019 and 2020, and this trend is likely to be maintained. New research suggests that heatwaves that trounce previous records by 5°C or more will be up to seven times more likely in the next 30 years, and up to 21 times more likely from the middle of the century onwards. The stark message is: 'You ain't seen nuthin' yet.'

Those impacted most, in the years ahead, will be people without access to air conditioning: many in low-income countries, but millions more in the industrialised nations. Worst affected will be those living in cities where the 'urban heat island effect', sustained by the expanses of concrete and asphalt, pollution, lack of vegetation and waste heat from human activities, can amplify temperatures by several degrees.

Many, if not most, of the 70,000 deaths arising from the baking summer of 2003 occurred in cities, where the old and vulnerable were unable to escape the heat, and where night-time temperatures in some places did not fall below 24°C. Exposure to deadly heat in cities has tripled since the 1980s and almost one-quarter of the world's population

(1.7 billion people) is at risk. As migration from rural areas to cities continues in many countries, and temperatures continue to climb, this is a figure that is only going one way. And for most people there will be no escape. Costly air conditioning will not be an option, and most of the planet's homes are simply not built to handle the heat that future decades will bring – in both developing and developed countries. Huge numbers of modern, tiny, poorly insulated UK homes, for example, will become unliveable heat traps, responsible for thousands of deaths every summer by 2050. And despite repeated warnings, hundreds of thousands of these inappropriate homes continue to be built every year.

In many parts of the world, summer is already becoming a time to dread rather than embrace. In Australia, many residents no longer look forward to a time of year that now brings unbearable heat, widespread drought and deadly wildfires. In India, pre-monsoon heat that threatens to smash the 50°C mark is anticipated with growing concern. In California, they are planning to name heatwaves as they do major storms, so that people will take more notice and make preparations. Maybe because we feel we are starved of sun for the rest of the year, summer heat still seems to be celebrated by many Brits. As temperatures go on to top the 40°C mark, and with heat-related deaths across the country set to exceed 7,000 a year by the middle of the century, this is an attitude that will inevitably change.

The heat that kills in hours

Imagine the hottest you have ever felt, then double it. Imagine heat so unbearable that you can barely breathe or lift a finger; that, despite burning up inside, you are unable to sweat so your body cannot cool down. Imagine knowing that – unless you can find an air-conditioned refuge – you have just hours to live.

No, this isn't a description of what it feels like in the world's hottest sauna, it is a foretaste of the horror of humid heatwaves that hundreds of millions may face later this century. In fact, such inescapable heat and humidity is nothing less than the logical end product of the increasingly extreme summers that are now rampant across the planet.

The thing is, all heat is not the same. We are probably all familiar with the fact that a combination of high temperatures and elevated humidity makes it feel far hotter than high temperatures on their own. The record-breaking temperatures, sometimes in excess of 50°C, that have made the news in recent years, are all measured using a standard – or 'dry-bulb' – thermometer. As such, they provide an accurate measure of the air temperature, but say nothing about the prevailing humidity, so they don't provide any information about how hot it felt to the people on the ground.

In addition to the dry-bulb temperature, meteorologists also measure the so-called wet-bulb temperature (WBT), which provides a measure of heat and humidity combined and is a far more accurate metric for what people actually feel. The WBT is measured on a wet-bulb thermometer that is in essence an ordinary thermometer encased in a wet cloth. Because heat is absorbed by the water and carried off in passing air, the WBT is almost always lower than that measured on a normal dry-bulb thermometer.

The closer the WBT gets to the dry-bulb temperature, the higher the humidity. If the temperature on the two thermometers is the same, then the humidity has reached 100 per cent. The key take-away is that, at this level of humidity, sweat will not evaporate into the air, which is already saturated with water, so it will no longer work to keep the human body cool.

Air temperature and humidity are combined in a 'heat index' that provides a measure of the 'felt air temperature'. In other words, how hot you would feel, in the shade, under those conditions. Even people used to heat, for example, agricultural workers in the tropics, would not be able to carry out normal activities at a WBT of 31°C (88°F), which would actually feel like 50°C (122°F). The US National Weather Service regards a WBT of 31°C as representing extreme danger to humans and warns that the failure to take immediate precautions could result in serious illness or even death. The latest dire forecasts predict that with a 2°C global average temperature rise, more than 1 billion people will be affected by such extreme heat stress – this figure rising to fully half the world's population with a rise of 4°C.

Things start to get even nastier at a WBT of 35°C (95°F), which would correspond to a heat index temperature of a staggering 70°C (158°F), although the index does not actually go this high, not yet at least. This flags the onset of a heat and humidity combo at which the human body can no longer cool itself down through sweating. This means that someone in the shade, even if they are young and fit, even with access to as much water as they want, would be unlikely to survive for more than six hours or so. Normal body temperature is 37°C, but if the body is unable to lose heat, then this core temperature will just keep rising, ultimately resulting in organ failure and death.

Climate models have forecast that the first WBT 35°C humid heatwaves will not happen until the middle of the century, at the earliest. Recent research, however, has revealed that these conditions have already been achieved in the Persian Gulf. So far, such extreme heat and humidity has been localised, and has only persisted for an hour or two at a time, but many other parts of the subtropics have experienced conditions very close to the 35°C survivability limit. Consequently, coming decades bring the prospect of such deadly heat prevailing at a regional scale and lasting for much longer, resulting in large parts of the subtropics reaching the very limit of liveability for humans.

Without large and rapid cuts to greenhouse gas emissions, more and more of the planet is forecast to come under severe threat from such malignant heat. The Persian Gulf, South and South-east Asia and China seem likely to bear the brunt. The Indian subcontinent will be hit particularly hard too. Sometime during the later decades of the century, millions of people, including in the major cities of Lucknow (Uttar Pradesh) and Patna (Bihar) are predicted to face at least one WBT 35°C heatwave, bringing the prospect of massive death tolls. Furthermore, more than 1.2 billion people will be exposed to WBT 31°C heatwaves, especially those farming in the Indus and Ganges valleys.

Ground zero, however, looks like being China's northern plain, the country's agricultural heartland where, today, 400 million people toil in already debilitating summer heat. In the absence of serious emissions cuts, fatal humid heatwaves are forecast to strike the region repeatedly after 2070. Without widespread and immediate access to air-conditioned sanctuaries, this would effectively make China's breadbasket uninhabitable, at least during the hottest months.

A number of major population centres in, or close to, the region will also be affected, including China's biggest city,

Shanghai. Here, WBT 31°C heatwaves are forecast to flourish hundreds of times between 2070 and the end of the century, with the deadly WBT 35°C threshold potentially being crossed on about five occasions. This will simply be continuing a trend that has seen the number of extreme heatwaves across the northern plain rising substantially over the past 50 years, culminating in 2013 in unprecedented drought and heatwave conditions that persisted for almost two months.

It is an especially disturbing thought that, if global heating continues unchecked, ever larger parts of the world that brought forth and shaped our species will, for the first time, become off limits to us.

Scorched Earth

In July 2018, much of Southern Europe was baking in a record-breaking heatwave that spawned wildfires right across the region. On 23 July, fires ignited in the Attica region of Greece, east of the capital, Athens. Spurred on by gale-force winds, the fires expanded into the community of Mati faster than residents could react, burning some people alive in their homes and cars and others as they huddled together within metres of the sanctuary of the sea. More than 100 people died, while many more needed to be evacuated from the beaches. This was the second most lethal wildfire of the 21st century, but the years since have brought an explosion of wildfire activity, so it is unlikely to be a record held for long.

Wildfires are emblematic of our rapidly heating world and, in recent years, have become ubiquitous features of both northern and southern hemisphere summers. So much so that the geographical spread of deadly wildfires has expanded as quickly as the fires themselves. Previously viewed as a preserve mainly of California and Australia – where the devastating 2019/20 wildfire season saw 6,000 buildings obliterated and left 34 people dead – not a year goes by now without major conflagrations springing up in country after country, from the Siberian wastes of Russia to Brazil's Amazon rainforest. And their destructive power is increasing apace too, so that fires are not just destroying buildings but erasing entire communities.

Global heating is driving the increase in wildfire activity via longer and more intense heatwaves and droughts, which leave vegetation tinder-dry and at risk of ignition from lighting strikes or human activity. When they become established, wildfires can be self-sustaining. They do this by creating their own weather, promoting giant 'pyrocumulonimbus' clouds that spawn lightning strikes, triggering more fires. In the larger conflagrations, devastating fire tornadoes can form too, which rampage across the landscape bringing wind speeds in excess of 200 km/h.

Once the conditions are right, a major fire can be ignited by something as small as a spark from a train wheel or the magnification of sunlight through a piece of broken glass. Wildfires can start within seconds but often take weeks or months to bring under control. Once on the move, fires travel quickest uphill, where the rising flames act to pre-heat the vegetation higher up, making it easier to ignite.

In recent decades, global heating has been promoting bigger and more intense fires that are increasingly difficult to contain, and which may burn out of control for long periods. No previous fire season, however, has been able to hold a candle to that of 2021, which saw colossal blazes rage across western North America, Siberia, Greece, Italy and elsewhere in Southern Europe, Turkey, South Africa, Algeria, Israel and India. In California alone, more than 8,000 fires burnt across more than 10,000 square kilometres – an area equal to Lebanon – destroying more than 3,500 buildings. A major casualty was the historic gold rush town of Greenville, obliterated by the Dixie fire – the largest in the history of the state – which took more than three months to bring under control.

In the context of global heating, wildfires provide a twin boost. The first due to the destruction of huge areas of forest that would otherwise have absorbed a significant amount of carbon during its lifetime. The second because the burning of wood on such a prodigious scale releases colossal amounts of carbon dioxide. The biggest and most intense fires also reduce the density and size of trees that grow back, making them less effective at absorbing carbon.

In the seven months to the end of July 2021, the burning of forest and other vegetation, either intentionally or accidentally, had added a record 343 million tonnes of carbon dioxide to the atmosphere. More than half of this came from the North American and Siberian fires, smoke from the latter bringing serious air pollution to cities such as Yakutsk and drifting as far as the North Pole. Such high latitude fires, which are also becoming increasingly prevalent across northern Canada and Alaska, are especially concerning, as they can continue to smoulder in the peaty soils during the winter months before bursting into life again the following summer. For the whole of 2021, wildfires released around the same amount of carbon dioxide as Germany emits in a year.

Not all wildfires are accidental. Some are started with malicious intent, others deliberately to burn primary forest prior to cutting and clearing it for farmland. In 2019 alone, almost 4 million hectares of primary tropical rainforest was lost, the equivalent of a football pitch for every second of the year. As a consequence, another 1.8 billion tonnes of carbon dioxide was added to the atmosphere, an amount equal to the annual emissions from almost half of all the world's cars. In Brazil's Amazon rainforest, between August 2020 and July 2021, an area bigger than Cyprus was destroyed, much of it by intentional burning.

Wildfires that are now obliterating entire towns are also threatening larger communities. In 2017, fires drove many from their homes in Ventura County, close by Los Angeles, while in 2020 wildfires raged close to the outskirts of Sydney. August 2021 saw the flames lapping at the edge of California's South Lake Tahoe, inciting a sudden and chaotic evacuation of the 20,000 population. It can only be a matter of time before wildfires begin to plough into city suburbs, threatening many thousands of properties and lives.

It has become clear that fighting the huge new fires spawned by global heating is not working and containment is becoming increasingly difficult. Looking ahead, fire experts recommend that efforts need to be focused more on prevention, for example, by means of deliberate, cool-season, burns designed to destroy much of the old wood, leaves and brush that form the raw material for major wildfires.

As climate breakdown progresses, the threat of major destructive wildfires is forecast to increase dramatically in previously unaffected countries. As summer temperatures and drought conditions become more common in the UK, for example, the number of days every year with a very high risk of fire is likely to quadruple to more than 120 in the drier east.

In the western United States, the frequency of 'fire

weather' – the hot, dry and windy conditions that fuel big fires – has increased dramatically in the past 50 years. This is a trend that is certain to continue across the world as well as in North America. As the heat continues to build globally, images of destructive wildfires will become even more commonplace on our TV screens and news feeds.

METEOROLOGICAL MAYHEM AND SOCIETY ON THE EDGE

5

Rain, rain, go away

Everything is big in China, and that includes the floods. In July 2021, the central province of Hunan experienced astonishing levels of rainfall that triggered widespread, devastating flooding. Rainfall records were smashed across the region as more than 64 centimetres – a year's worth of rain – fell over a period of just 24 hours at rates which peaked at more than 20 centimetres an hour. Metro stations were deluged, dozens of reservoirs overflowed and nearly 1 million people were evacuated. More than 300 people died and 400,000 cars were destroyed or battered in the rapidly flowing floodwaters. The proximity of megacities to its great rivers means that China is no stranger to deadly, high-impact floods, but global heating is ensuring the country is now facing what used to be ruinous one in 1,000-year flooding almost every year.

China was not alone. In the same year, the worst floods and landslides in more than 100 years took many lives in India and Nepal, while in Japan torrential rains led to calls for almost 1.5 million residents to immediately leave their homes and seek safety. The unprecedented floods that took close to 250 lives in western Germany and Belgium were matched by equally ferocious flash flooding in Turkey.

None of this should really come as a surprise because a hotter world is also a wetter world. Warm air contains a higher proportion of water molecules in the vapour state, and the amount of water vapour in the atmosphere has already climbed by around 7 per cent since Arkwright's day. Furthermore, this trend will continue until global heating is reined in. And there is more water available too. As global heating progresses, higher temperatures push up rates of evaporation from the surface of the planet. Rates of 'evapo-transpiration', which is a combination of evaporation from the surface and transpiration from plants, are up 10 per cent on 2003 levels and still climbing.

As we say, and gravity decrees, what goes up must come down, so more atmospheric water translates directly into added precipitation. For every degree Celsius temperature rise, global mean precipitation goes up by somewhere between 1 and 3 per cent. Locally, however, increases can be much greater and would rise, for example, by more than 10 per cent in the tropics. If this additional rainfall was spread out evenly across the world and over time it would be less of a problem. The reality is, however, that rain is now falling in shorter, more intense bursts, which provide the ideal conditions for flood development. In particular, torrential downpours are conducive to flash flooding, whereby water overwhelms natural and artificial drainage systems and flows across the surface with little penetrating the soil.

One serious concern, looking ahead, is that climate breakdown looks as if it will exacerbate weather phenomena known as atmospheric rivers. These are long, narrow belts of air that together carry more than 90 per cent of the moisture from the tropics into temperate and polar regions. Each 'river' acts as a conveyor belt capable of transporting a greater flux of water than the Amazon, which can bring very heavy rain to a particular region that can last for days on end, saturating the ground and driving dangerous flooding. Atmospheric rivers are common enough climate features, with perhaps a dozen or so active across the planet at any one time. Modelling has demonstrated, however, that while global heating might actually reduce the number of such rivers, it will also result in atmospheric rivers becoming longer and wider. As things stand, the global frequency of atmospheric river *conditions* – which means persistent very heavy rain accompanied by strong winds – looks like being 50 per cent more frequent by the century's end.

More atmospheric rivers is the last thing we need, as residents of the UK's Lake District would no doubt agree. In both 2009 and 2015, atmospheric rivers brought record rainfall to the region. In 2009, the town of Cockermouth suffered major flooding and infrastructure damage, including the destruction of four bridges. Six years later, another river conspired with Storm Desmond to dump a record-breaking 34 centimetres of rain on the region, shattering the all-time 24-hour rainfall record for the UK.

Extreme rainfall events that happened once every ten years in Arkwright's time now occur every 7.5 years, and this frequency will reduce further to every three or four years if the world continues to follow a business-as-usual emissions pathway. A foretaste of the future was provided by severe flash flooding in New York in September 2021, which involved the city authorities issuing their first ever flash flood emergency warning. Subway lines were swamped in Manhattan, Brooklyn and New Jersey, while several people drowned as their basement apartments flooded too rapidly to allow escape. In the UK, towns and cities have been inundated by flash floods on more than 50 occasions over the past decade and a half alone, and this is only going to get worse. A report published early in 2022 flagged London as being especially susceptible and warned that there was significant risk of drowning as the capital's Victorian drainage system struggles to cope with the growing flash flood hazard in the years to come.

As extreme rainfall becomes ever more common, so increasing numbers of people are coming under serious threat. Between 2000 and 2018, somewhere between 255 and 290 million people worldwide were impacted by major flood events. Because population growth in some of the most flood-prone areas is higher than the global average, many more people will be at risk as the world continues to heat up and get wetter. One estimate suggests that ten times as many people will be threatened by floods over the course of the coming decade than at the turn of the century.

Looking ahead, then, increasingly destructive and deadly flooding is set to become one of the most obvious signatures of global heating. In Europe, the most dangerous slow-moving storms, such as the one that caused the 2021 German floods, are expected to be up to fourteen times more frequent by 2100.

In the UK, both winters and summers are already around 12 per cent wetter than over the period 1961–1990, and the amount of rain falling on extremely wet days has climbed by 17 per cent compared to the same period. Without significant emissions cuts, winters are set to be almost a third wetter by 2070, and summers close to 50 per cent drier. The latter figure is a little misleading as much of the reduced summer precipitation will take the form of torrential downpours during thunderstorms, raising the potential for severe flash flood events in the hotter months.

At the time of writing in late 2021, another atmospheric river barrelled into western Canada, bringing catastrophic flooding and effectively cutting off the city of Vancouver – a timely warning of what is in store for many in the decades ahead. As we shall see next, however, the problem for some parts of the world in the decades to come will not be too much rain, but far too little.

Dustbowl

In 1935, the construction of the Hoover Dam across the Colorado River resulted in the formation of Lake Mead, the biggest reservoir in the United States. Straddling the Nevada–Arizona border, the lake and the dam are critical elements of US infrastructure, supplying water and power to tens of millions of people across the states of California, Arizona and Nevada. But now there's a problem. A prolonged mega-drought that has plagued the region since 2000 has meant that the level of Lake Mead has been in continuous decline for more than twenty years and is now at a historic low. As a consequence, the dam has struggled to produce electricity, while water rationing has been introduced in the region for the first time.

The severe drought conditions affecting Lake Mead are the worst in more than 1,200 years and stretch across three-quarters of the western United States. The Colorado

River system itself, upon which one in ten Americans depend for at least some of their water, is now flowing at just half its normal capacity.

What is happening in the south-western United States is a reminder that, while it is true that a hotter world is also a wetter one, this is a general rule that does not apply everywhere. Higher temperatures, unsurprisingly, also drive an increase in arid conditions so that extreme, prolonged drought and the expansion of deserts are only to be expected. And this trend is now becoming apparent in many parts of the world. In 2020, almost one-fifth of the global land area was afflicted by drought in any single month, a figure that never exceeded 13 per cent in the second half of the 20th century.

Since 2014, Europe has experienced the most extreme drought conditions in more than 2,000 years, wiping out crops and setting the scene for the worst wildfires on record. In 2020, extreme drought affected much of South America, massively impacting agriculture and causing losses of close to $3 billion in Brazil alone. Across the Pacific, meanwhile, Australia's long-standing drought conditions have spawned frequent wildfires that have decimated the sub-alpine forests of Victoria and New South Wales. In Asia, in 2021, Taiwan's reservoirs were effectively empty, while at the same time, in the Central Asian republics and parts of Russia, unforgiving drought conditions resulted in widespread livestock deaths and shortages in animal feed.

Drought may not be as spectacular or powerful as other meteorological phenomena, but it should never be underestimated. It is, after all, regarded by cultural historians as one of the most important drivers of political change in the past 5,000 years of human history. Already this century, more than 1.5 billion people have been afflicted by droughts that have cost economies at least $124 billion. Indeed, the United Nations (UN) has warned that drought is a hidden global crisis that threatens to become 'the next pandemic' in terms of its potential impact on global society and economy. By the end of the century, the organisation forecasts that 129 countries will see an increase in exposure to drought, mainly due to climate breakdown, and a further 38 as a consequence of a conspiracy of climate breakdown and population growth. That is getting on for nine out of ten of all the nations on the planet being affected by the phenomenon.

In a taste of things to come, California's Lake Oroville hydro-electric power plant, which provides electricity for nearly 1 million people, was forced to shut down in August 2021 by drought conditions that had reduced the lake level by three-quarters. Another 100-foot fall in the level of Lake Mead and the Hoover Dam power plant could follow suit, resulting in a drought-driven power crisis that could have catastrophic consequences for manufacturing and food production, especially in California, including Silicon Valley.

Similar threats exist across the planet wherever nations or regions are dependent on hydropower for energy. In autumn 2021, Brazil's government demanded a 20 per cent cut in electricity usage from consumers and businesses, as failed rains and drought meant that many dams were working at reduced capacity. Two-thirds of Brazil's electricity is provided by hydropower, making the country especially vulnerable as droughts become more intense and longer lasting.

The impacts of drought on power production is particularly severe across Africa, with reduced reservoir levels causing power shutdowns or reductions in Malawi, Tanzania,

Mozambique and Zambia in recent years. The fact that many African countries are almost entirely reliant on hydropower is a recipe for disaster, as high temperatures and ever more severe and sustained drought conditions will mean that power production in the decades ahead is likely to become increasingly compromised.

Of course, drought isn't simply a threat to power supplies; it also brings a critical shortage of water to drink and typically has a devastating impact on agriculture, often leaving deadly famine in its wake. In 2021, water shortages are estimated to have affected more than 3 billion people, and as global heating bites ever harder this situation is unlikely to improve. In some places, such as sub-Saharan Africa, severe drought routinely devastates crops and pastureland every three years or so, making life for 50 million people precarious, to say the least. With four-fifths of all the world's cultivated land dependent upon regular rainfall, future widespread drought has the potential to have catastrophic consequences for global food supplies.

As is happening in the south-western United States, land can become locked into an inescapable cycle of extreme heat, drought and wildfire, ultimately leading to unliveable conditions that drive outward migration. This is something that will become far more common in a hotter world, notably in those areas where drought conditions are predicted to become ever more severe, including Southern Europe, Australia, Central Asia, the United States, Central and South America and most of Africa. The end game will be the transformation of formerly usable land to desert. Almost 2 billion people live in regions vulnerable to desertification, which could add another 50 million people to the tally of the displaced by the middle of the century.

Storm force

It's going back a bit now, and there is no evidence to explicitly link it to global heating, but it's worth revisiting the great storm that struck the southern UK in October 1987, if only to provide a feel for what the future is likely to bring. I remember the occasion especially clearly as I went into a pub for a drink on a chilly autumn evening and came out into a balmy night more reminiscent of high summer. In fact, as the storm approached, temperatures climbed by a mind-boggling 10–12°C in less than half an hour. And, of course, along with the temperature hike came the wind. The maximum gust speed of 185 km/h (115 mph) was registered at Shoreham-by-Sea on the south coast, but gusts exceeded 160 km/h (99 mph) in many places. Sustained wind speeds over parts of southern England reached 121 km/h (75 mph), which is comparable to a category one hurricane in the tropics. In which case, it is no surprise that the country took a real battering. There was widespread structural damage and countless roads and railways were closed by debris and fallen trees. Several hundred thousand homes were left without power and 15 million trees were brought down.

The Great Storm of 1987 – named 87J by the insurance industry that sucked up the huge financial losses – was the most powerful storm since 1703 and supposed to be a one in 200-year event. Yet, just three years on, a comparable tempest, known as the Burns' Day Storm, gave the UK a second beating and continued eastwards to wreak havoc across Europe. Wind speeds again exceeded 160 km/h (99 mph) in many places, causing extensive structural damage and severe flooding and taking 47 lives.

Storms such as 87J, which assail the UK, the adjacent

continent and other mid-latitude land masses on a regular basis, are termed extratropical cyclones, to distinguish
them from their tropical cyclone siblings. As yet, there is no
clear evidence that overall storminess at mid-latitudes has
increased as global heating has progressed. Looking ahead,
however, the picture is far from rosy. Across the UK and
Europe, for example, annual losses from windstorms average
just north of $2 billion, but this figure will rise significantly
as global heating is slated to launch more frequent and more
powerful extratropical cyclones at the region.

And there is another threat too. As the planet continues
to heat up, warmer oceans at mid-latitudes are predicted to
increase the number of hurricane-like storms from the tropics landing in our part of the world. At the moment, such
events are rare, and Hurricane Ophelia in 2017 is one of the
few examples. Ophelia grew out of a tropical storm in the
eastern Atlantic that built to a powerful category three hurricane close to the Azores. She then followed a highly unusual
track, heading north-east towards the Iberian Peninsula
instead of westwards towards the United States. Ophelia
weakened as she approached Europe, but still retained all
the features of a hurricane, notably an 'eye' surrounded by
a spiral of clouds. After turning northwards, she eventually
churned across Ireland, bringing gusts of nearly 200 km/h
(124 mph) that caused serious damage and widespread
power outages. For the next couple of days, Ophelia plagued
the UK and Scandinavia before slowly dwindling away.

It is not unusual for remnants of hurricanes to blow
themselves out on the shores of the UK and Europe, but
Ophelia remained a true hurricane for longer than any other
major storm headed our way. As hurricanes feed off ocean
heat, it seems inevitable that as the sea warms at higher

latitudes, hurricanes will be able to edge ever further northwards. The worry is that more and more will follow a similar track to Ophelia and make it to the UK and Europe while still maintaining hurricane characteristics and wind speeds. The consequences could be devastating and lead to widespread damage and fatalities across the region.

Meanwhile, in warmer climes, global heating is already beginning to cause increased mayhem on the storm front. Tropical cyclones – an overarching term that includes Atlantic hurricanes and Pacific typhoons – are already morphing to become more deadly and destructive. While the jury is still out on whether global heating is increasing tropical cyclone frequency, there is little doubt that they are becoming more powerful. Globally, the proportion of storms allocated 'major' status – that is category three to five storms (on a scale of one to five) with sustained winds of more than 178 km/h (111 mph) – has increased significantly, a trend that is set to continue. This is particularly concerning as these storms cause 85 per cent of all tropical cyclone damage.

Looking ahead, the number of major tropical cyclones is forecast to be up by as much as 25 per cent globally by the period 2081–2100, and more than 30 per cent higher in the Atlantic. Worse still, the comparable numbers for just the most severe category five storms – which have sustained wind speeds of more than 252 km/h (157 mph) – are 85 per cent and 136 per cent respectively. Some have suggested that by the early 22nd century, some tropical cyclones may have become so powerful that a new category six may need to be added at the top end of the scale.

It goes without saying that such storms would be capable of unimaginable damage, only a taste of which was provided by the arrival of Hurricane Andrew on the east coast of Florida in 1992 – one of just three category five storms to make landfall on the US mainland. Wind gusts as high as 280 km/h (174 mph) stripped buildings from their foundations, destroying in excess of 60,000 homes and damaging 120,000 more. A hypothetical category six cyclone would unleash devastation on another level altogether.

Higher wind speeds also means bigger 'storm surges'. In 2005, a 9-metre surge associated with Hurricane Katrina swamped flood defences, inundated more than four-fifths of the city of New Orleans, and took 1,800 lives. Shocking as these numbers are, this event may prove to be insignificant in comparison to the devastating power of surges forged by the super-storms of the future.

On top of becoming more powerful, tropical cyclones are responding to rising global temperatures by holding more water, which means more rain. They are travelling at a slower pace too, even stalling at times, so that more rain falls in the same place, dramatically increasing the severity of flooding. As such, Hurricane Harvey, which dumped more than 1.5 metres of rain on parts of Texas in 2017, causing a mind-blowing $125 billion worth of flood damage, is simply a foretaste of what will become far more commonplace on a hotter planet.

Communities far away from the coast are also in the firing line, as a hotter atmosphere is expected to spawn more and perhaps bigger tornadoes. The exact relationship between global heating and tornadoes is complicated and yet to be fully understood, but changes do seem to be afoot. In the United States, the killing grounds of these spinning columns of solid wind, which make up so-called Tornado Alley, seem to be migrating further east and are becoming more common in the winter months. As such, events like the devastating Kentucky tornado of December 2021, which took more than 90 lives, will probably become increasingly common.

Food, famine and flare-ups

Nothing highlights the complex and wide-ranging ramifications that global heating and climate breakdown will inflict upon society more than the interrelationships between drought, food, famine, migration and conflict.

Inevitably, in a hotter world, drought will impact increasingly severely on agriculture, which in turn will translate into food shortages and famine in the majority world and price hikes in developed countries. In the longer term, sustained drought conditions will drive mass migration and measures to protect dwindling water supplies, bringing civil strife and cross-border conflict.

The African continent, in particular, has never been far from famine, but now global heating is acting as a threat magnifier, making bad situations even worse. In 2021, it was East Africa that bore the brunt. In southern Madagascar, a 10 per cent rise in temperature and a 10 per cent decrease in rainfall has provoked long-term drought conditions leaving more than 1 million people under threat of starvation. Many were forced to eat cactus leaves, termites, clay and ashes to stave off hunger, while the rate of child malnutrition doubled in just a few months. Dubbed by the UN as the first climate change famine, at the time of writing there is still no sign of things improving any time soon. Indeed, the situation has been made even worse by the arrival of three devastating cyclones in the early months of 2022.

In 2021, famine was also prevalent in Ethiopia, South

Sudan and Yemen, while in Kenya extreme drought affecting half the country put more than 2 million people at risk of starvation. Sustained drought conditions across much of East Africa have been exacerbated in recent years by a combination of severe flash flooding and plagues of locusts.

The UN World Food Programme estimated that 41 million people in 43 countries were 'teetering' on the edge of famine in 2021, up from 27 million in 2019. As global heating continues to accelerate and extreme weather becomes ever more common, this figure is only going one way.

Climate breakdown is not yet sufficiently advanced to bring the prospect of widespread starvation to developed countries, but in late 2021 it is already having an impact through commodity shortages and price hikes, as crops succumb to extreme weather. Most obvious is the sudden rise in pasta prices, reflecting the destruction of much of the Canadian durum wheat crop by extreme summer heat and drought. And if you like a nice glass of red to accompany your spaghetti Bolognese, you may have to spend more on that too. Vineyards across Europe took a real battering from extreme weather in 2021, resulting in wine production plummeting by 27 per cent in France and 25 per cent in Spain. Add to this rocketing coffee prices due to severe weather wiping out a third of Brazil's crop, and cocoa production in Africa hit by drought, and the news for dinner party hosts is not good.

And these are just the forerunners of the serious food supply issues that will arise in a hotter world. The bottom line is that some crops simply will not grow under the higher temperatures that are becoming the norm, many that can will see yields fall away and others will be lost to extreme weather. In this regard, the cherries and blueberries that baked where they grew and the apples that roasted on the trees during the 2021 North American heatwave should be regarded as early examples of the far more widespread crop damage to come. Bearing this out, a recent study forecast that by 2100 crop damage arising from extreme weather is set to increase as much as tenfold.

We don't, however, need to wait until the end of the century. Even a global average temperature rise of 1.5°C, which we are likely to see within a decade, will have a serious impact on food supplies. The big worry is that four-fifths of all calories consumed across the world come from just ten crop plants, including the staples, wheat, maize and rice. Yield decreases for these key crops will thus have a disproportionate and massive impact on calorie intake, particularly in majority world countries. Already, global maize production was down 6 per cent in 2020, winter wheat down 3 per cent and rice down by nearly 2 per cent, compared to the 1981–2010 average.

According to one study, climate breakdown is now reducing consumable calories from the ten main crop plants by around 1 per cent. This may not sound like much, but it represents a loss of 35 trillion calories a year, enough to feed more than 50 million people. By 2050, global production of maize is predicted to fall by almost a quarter, rice by 11 per cent and potatoes by 9 per cent. Wheat fares a little better, falling by just three percentage points, but together these figures point towards a massive reduction in consumable global calories, which will have a catastrophic impact, especially in majority world countries. And a footnote for those who can't do without their cuppa. The main tea-growing areas in the world are set to be especially badly hit by severe weather, resulting in massive yield falls. Kenya, the world's biggest producer of black tea, which supplies almost half of all the tea drunk in the UK, will see the area best suited to tea growing reduced by a quarter by the middle of the century, while areas that are marginally suited to production are likely to be cut by almost 40 per cent.

There is a suggestion that as some parts of the world become unsuited to growing a particular crop, others may become more amenable. This is plain wrong. For example, in coming decades, rising UK temperatures and reduced frosts may appear – at first glance – to make the country more suited to growing grapes. In reality, however, anyone who tries to do so will need to battle a cocktail of increasing extreme precipitation, flooding, heatwaves, drought and new pests. As far as agriculture is concerned, there are no winners in the global heating game. Everyone loses.

As heat, drought and crop failures drive people from their homes and off their land, the numbers of economic migrants are forecast to go through the roof. By 2050, more than 250 million people could be on the move across sub-Saharan Africa, South Asia, Latin America and elsewhere, having been forced on to the road by climate breakdown. Twenty years later, fully a fifth of the Earth's land area may be effectively uninhabitable, provoking even higher numbers of migration.

As large numbers of hungry and desperate people head for the cities and across borders, it is inevitable that civil disorder will become widespread and clashes between neighbouring nations almost commonplace. Developed countries, particularly the United States, the UK and those that make up the European Union will, without doubt, become target destinations of many of the uprooted, causing an explosion in trafficking and modern slavery, bringing serious security problems and, surely, promoting migration policy rethinks. I will explore the links between climate breakdown and food, famine, migration and conflict in more detail in due course.

GOING UNDER

6

Earth's ocean elevator

With more than four in ten people worldwide living on or within 150 kilometres of the coast, there is vested interest in knowing whether the sea is going up or down, and how quickly. After all, a single metre of sea level rise can doom the lives and livelihoods of 250 million people in coastal communities.

In all, 97 per cent of all the world's water – a colossal 1.3 billion cubic kilometres – is held in the giant repository that is the world's oceans. How high or low the upper surface of the ocean rests in relation to the surface of the land we live upon is determined, firstly, by how much water is locked away as ice at the poles and, secondly, by the width and depth of the ocean basins.

Throughout our planet's immensely long history, the level of the sea has yo-yoed dramatically in response to natural changes in the climate, which influences the amount of ice at the poles, and in the disposition of the continents, which affects the geometry of the ocean basins.

During ice ages, an extra 4 per cent of ocean water is sequestered and used to build the great continental ice sheets, so that sea level falls by 100 metres or more. Low sea levels also occur whenever the ocean basins are wide and deep, which tends to happen when the incrementally slow dance of the continents across the face of the Earth brings them together to form a single supercontinent. On these occasions, the sea *regresses* from the land, leaving more of the latter exposed.

At other times, for example, when the continents are spread far and wide – as they are today – the ocean basins are more limited in area and shallower. Then, ocean waters with nowhere else to go spill over and *transgress* across vast tracts of low-lying topography adjacent to the margins of the basins.

Our knowledge of what sea level was doing way back in the Precambrian, more than 541 million years ago, is pretty sketchy. However, studies of more recent ancient shorelines and the pattern of marine sediments have helped to paint a pretty clear picture of the highs and lows of the oceans.

A plot of the sea level switchback across the 500 million years or so defines a distinctive 'M' shape, with two peaks and three troughs. Following the break-up of the Pannotia supercontinent in the late Precambrian, sea levels rose to a peak around 450 million years ago, during the Ordovician Period, when they touched an astonishing 200 metres, or more, higher than they are today.

As the continents came together again to form Pangaea, so global sea levels fell substantially and, by 250 million years ago, in the Permian, they reached a low point comparable to that of today's oceans. The disruption of Pangaea saw levels rise again to about 170 metres above today's values, peaking in the late Cretaceous, around 80 million years ago. Since then, and notwithstanding the wild swings in sea level over the past couple of million years as glacial episodes alternated with interglacials, it has been pretty much downhill all the way.

At the height of the last ice age, around 20,000 years ago, global sea levels were up to 130 metres lower than today, exposing land that now lies far beneath the waves. The UK, for example, was connected to mainland Europe, and what is now the North Sea's Dogger Bank was inhabited. Elsewhere, land bridges linked Alaska and Siberia, the many islands of Indonesia, and Australia and Papua New Guinea, facilitating a diaspora of our ancestors to previously uninhabited parts of the planet.

As temperatures climbed and the great continental ice sheets wasted away, the depleted ocean basins were rapidly replenished. For much of the time, sea level rose slowly and steadily. On other occasions, however, colossal influxes of water from giant lakes of glacial meltwater pushed the level up by several metres in just a few centuries or even less. By around 8,000 years ago, sea level was pretty much at the same level as in Arkwright's day, where it remained stable until global heating began to take a hand.

The level of the sea in ancient times is determined using proxy methods, for example, the heights of old shorelines and the mapping of marine sediments and beach deposits. In modern times, it has become possible to measure sea level directly, and to track any changes, using tide gauges. Self-recording tide gauges using mechanical floats have been around in some form or another since 1830 and can still be found in some places today. Most gauges in use now, however, are more technically sophisticated and use pressure, acoustic or ultrasonic sensors to measure tidal variations and to detect changes in sea level over time.

The picture painted by bringing together tide-gauge records from across the world reveals that sea level is on the move again and has been for some time. Over the course of the 20th century, global sea level climbed by around 20 centimetres as global heating began to initiate the melting of mountain glaciers and ice at the poles and warmer oceans expanded and took up more space. For the first time in several thousand years or so, the world's oceans were on the rise and, uniquely in the long history of our planet, this was not driven by natural processes.

Tidal gauges still provide a useful tool for doing the job they were designed to do, which is essentially measuring and characterising tidal variations, but they are not really up to the job of tracking the tiny annual changes in sea level that mark the ocean's response to global heating. Fortuitously, there are now satellites that have this capability, and which can provide a planet-wide perspective on the pattern of sea level rise, which is not consistent across all the world's oceans. As we shall see next, the story they tell is a worrying one.

The seas rise up

In 2005, residents of Tegua, a tiny island in the South Pacific's Vanuatu archipelago, were forced by rising seas to abandon their coastal villages and move inland to a safer area. Still only a couple of metres above sea level, the new location was unlikely to provide a long-term solution, but it would do for the time being. Although disputed by some, the villagers of Tegua were recognised by the UN as the world's first climate change refugees, uprooted as a direct consequence of mankind's impact on the environment. If true, they certainly won't be the last.

For the inhabitants of Tegua and other low-lying Pacific islands, climate breakdown is a reality right now. Already, several islands – albeit uninhabited – have been permanently submerged, and many others, including the Marshall Islands, the Solomon Islands and Tuvalu, face a dismal future as the ocean continues to transgress ever further onto their limited land area. In the Indian Ocean, the Maldives is at particular risk, the average height of the land across its 1,100 islands just touching 1.2 metres above sea level. So too is Bangladesh, where a projected 65-centimetre sea level rise by the 2080s would see nearly half of all productive land in the south of the country vanish beneath the waves.

The bad news is that sea level ultimately increases by 2.3 metres for every 1°C rise in global average temperature. This means that even if we could limit this rise to 1.5°C, sea level would still end up – in due course – more than 3 metres higher. Outlandish and scary as this sounds, it fits with conditions during the previous interglacial period when the climate was comparable or a little warmer than today and the sea level several metres higher.

The only outstanding question is how long a multi-metre rise would take. In its sixth assessment report, the IPCC forecast that, by 2100, sea level will most likely be around a metre higher than it is today, which is sufficient to affect the lives of hundreds of millions of people. But the true picture could easily be far worse. As the IPCC acknowledges, and a number of other studies support, it is perfectly possible that the rise in sea level by the century's end could be double this, maybe even more.

Observations of what sea level is actually doing also suggest that the IPCC is underplaying the threat. Between 1900 and 1990, global average sea level rose by 1.4 millimetres per year, hardly sufficient to be concerned about. But look at what's been happening over the past 50 years or so. For the period between 1970 and 2015, the figure was 2.1 millimetres per year and between 2006 and 2015, it was 3.6 millimetres per year. Between 2015 and 2019, the annual rate of global average sea level rise reached 5 millimetres, and it is unlikely to stop there. Bear in mind that for every centimetre rise in sea level an additional 6 million people come under threat from coastal flooding, and the urgency of the situation becomes pretty clear. Most worrying is the serious possibility that sea level rise might now be following a path that sees the rate doubling every twenty years or even less. A doubling period of twenty years would mean that by 2040 we could expect it to be rising by a full centimetre a year, increasing to 2 centimetres by 2060, 4 centimetres by 2080 and 8 centimetres annually by the century's end.

The main driver behind the rapidly accelerating sea level rise is a matching increase in the rate of melting ice. This is ramping up across the planet, everywhere from mountain glaciers to small ice caps, to the great polar ice sheets. By far the majority of the rise is coming from escalating rates of melting of the Greenland and West Antarctic Ice Sheets, about which more in the following section.

Whether or not the worst-case projections for sea level rise by 2100 come to pass, the crux of the matter is that we know now that we are already committed, however long it takes, to a rise big enough to swamp the world's coastlines and most of the biggest cities on the planet. By the middle of the century, rising seas are predicted to cause high tides to encroach upon land currently occupied by around 150 million people, making living there essentially untenable. This figure includes 30 million people living in China, one of the worst affected nations due to its high coastal population density. Assuming just a 1-metre sea level rise by the end of the century, more than 400 million people could then be living on land that is less than 2 metres above sea level, and hugely vulnerable to future rises.

A key point in understanding the threat from rising sea level is that it is not the same everywhere, nor for everyone. Factors including whether the land is rising or sinking and the influence of ocean currents mean that it varies from place to place. While the average stretch of coastline, therefore, experiences a sea level rise of around 3 millimetres per year, a coastal resident, on average, is faced with an annual sea level rise of 8–10 millimetres – pretty much double the average rate. This is because many live in cities built on river deltas, for example, Shanghai, Jakarta and New Orleans, which are subsiding due to water extraction from the underlying sediment. Jakarta, the Indonesian capital, is sinking at 10 centimetres a year – twenty times the rate of global average sea level rise, while New Orleans has already sunk so much that half of it is now below sea level and only protected by sea walls and levees.

Such dense population centres are clearly most at risk from eventual permanent inundation by the ocean. They are also increasingly under threat from storm surges and coastal flooding, the impacts of which will be greatly magnified as sea level continues to climb. So-called extreme sea level events will become increasingly common as the world continues to heat up, so that as early as 2040 what was a one in 100-year coastal flood could happen every year. On top of this, infiltrating salt water at higher sea levels can damage crops and contaminate supplies of fresh water. It can even rot the concrete foundations of buildings, causing damage and eventual structural failure. Indeed, it has been suggested that this may have been a factor in the 2021 collapse of an apartment block in the Florida town of Surfside, which took almost 100 lives.

Here in the UK, the Thames Flood Barrier was closed in October 2021, for the 200th time in its history, to protect London from the sea and from river flooding due to heavy rains. Although touted by the UK Environment Agency as being able to do the job until 2070, the barrier will become increasingly ineffective as sea level climbs and more intense rainfall drives greater river flow. Meanwhile, many other parts of the UK coastline have little or no protection. A scary reminder of how vulnerable the UK is to rising sea levels is provided by the fact that just a 2.5-metre rise – perfectly possible by the end of the century – would present an existential threat to Boston and Spalding and transform both Peterborough and Cambridge into seaside towns.

Goodbye Greenland?

After Antarctica, the Greenland Ice Sheet is the greatest body of frozen water on the planet, with a volume of more than 2.8 million cubic kilometres. The average thickness of the ice sheet is more than 2 kilometres, and greater than 3 kilometres in places, and its extent almost the size of Mexico. The critical statistic, which should concern us most, however, is this: if all the Greenland Ice Sheet melts, global sea level will climb by around 7 metres.

Greenland has hosted major ice cover for a very long time, since the middle of the Miocene Period, 14–11 million years ago, in fact. Since then, it has waxed and waned somewhat, but has always been present in some form or other. Dating of the deepest ice layers encountered in boreholes suggests that the ice sheet has been stable for the past million years or thereabouts. Recent research has demonstrated, however, that almost all of the ice sheet appears to have melted at least once in the 1.5 million years before that. This is not good news, as it points to the ice sheet being less stable than previously thought and increases the potential for current global heating to drive wholesale melting and ice sheet collapse.

During the relatively stable climate of the past several thousand years, the volume of the Greenland Ice Sheet has remained largely unchanged, but not anymore. Over the past few decades, the ice sheet has begun to fall apart at an ever-greater rate. As temperatures across the region climb, the surface of the ice sheet has succumbed to increased surface melting, mainly in summer. Meltwater collects to form lakes, some of which have volumes of tens of billions of gallons. Some of the lakes feed streams that travel across the surface of the ice to the coast. Others, however, drain into the ice sheet interior via fissures and crevasses. The meltwater then travels along the base of the ice sheet, where it comes into contact with the bedrock, until it reaches the sea. At the same time, so-called tidewater glaciers, which push out over the sea surface, are also being weakened by submarine melting due to a warmer ocean, which hastens their break-up – a process known as calving.

In the 1990s, the ice sheet was losing 33 billion tonnes of ice a year. During the past decade, however, the rate of ice loss has been *seven times* greater. Between 2002 and 2016, there was a net ice loss of 260 billion tonnes a year, and for 2018–2019 alone it was a colossal 329 billion tonnes. In total, over the period 1992–2018, the ice sheet shed an astonishing 3.8 trillion tonnes of ice. Such prodigious numbers are difficult to comprehend, so let's look at things another way. Consider the amount of water held in seven average-sized public swimming pools, then imagine that amount of water pouring off the Greenland Ice Sheet and into the North Atlantic for every second of every day of every year. If that doesn't scare you, it should.

It should come as no real surprise that the Greenland Ice Sheet did not fare well during the sweltering summer of 2021. In July, the temperature across parts of the ice sheet reached an extraordinary 19.8°C, leading to the loss of almost 17 billion tonnes of ice over the course of just two days. During the month of July as a whole, melting was happening across nearly two-thirds of the ice sheet surface. There were also record losses of ice due to the calving of glaciers where they meet the sea, launching fleets of giant bergs out into the Arctic Ocean. Submarine melting, due to the eroding effect of warmer seawater on submerged ice, hit record levels too. And, to top it all, on 14 August, rain was recorded, for the first time, falling at the summit of the ice sheet. Here, 3,216 metres above sea level, the annual average temperature is a frigid –30°C, so the arrival of rain provides the most shocking evidence yet that the Greenland climate is unequivocally entering completely new territory.

What happens next to the Greenland Ice Sheet is one of the great unknowns of climate breakdown science. As the global average temperature continues to rise, there is no reason to think that the rate of melting will slow down, and every reason to assume that it will continue to accelerate, contributing ever more to rising sea levels. The big fear, as mentioned previously, is that a tipping point will be reached, beyond which the collapse of all – or most – of the ice sheet is a done deal.

Research unveiled in 2021 suggested that we were on the brink of a tipping point that would see unavoidable melting of part, but not all, of the Greenland Ice Sheet. According to the study, the giant Jakobshavn glacier, which drains almost 7 per cent of the entire ice sheet, looked as if it was becoming locked into a positive feedback loop. Surface melting, it appears, is acting to reduce the height of the glacier, bringing it into contact with warmer air at lower altitudes. This, in turn, results in further melting that lowers its height further, and so on. If the Jakobshavn glacier has crossed its tipping point, or is about to, then this will lock in a 1- to 2-metre sea level rise.

The reality is, no one really knows if we have passed the Greenland Ice Sheet tipping point, partial or not. And even if and when we are sure, there are no real constraints on how quickly the collapse of the ice sheet will occur. Current convention has it that it would take thousands of years for the entire mass of ice to vanish. Some researchers, however, think break-up could happen far more quickly. It is especially worrying that very rapid sea level rises during the previous interglacial, of as much as 3 metres a century, may have been driven by sudden large-scale melting of the Greenland Ice Sheet.

When and if Greenland begins to lose its ice in earnest, a cascade of potentially catastrophic additional effects – on top of rising sea levels – could be triggered. Scariest of these is the possibility that the deluge of freezing water into the North Atlantic will bring the Gulf Stream and associated currents to a grinding halt. In addition to driving rapid regional cooling this would also result in major disruption to global weather patterns.

Dispatches from down under

While the Greenland stats are impressive, those of Antarctica knock them into a cocked hat. Almost two-thirds of all the planet's freshwater is locked away in the ice that covers 98 per cent of the Antarctic continent – a mind-blowing 26.5 thousand trillion tonnes of the stuff. The Antarctic Ice Sheet has been around quite a bit longer than its northern counterpart, having begun to develop as far back as the middle of the Eocene Period, 45.5 million years ago. Since then, it may have melted, on and off, around the margins but the core has remained ever present. Should all the Antarctic ice ever melt, then global sea level would rise by around 58 metres. Needless to say, this would be a cataclysmic event for human civilisation. We can, however, let out a small sigh of relief as no climate breakdown model – yet, at least – suggests that global heating is likely to bring this about.

Nonetheless, the Antarctic Ice Sheet is already suffering considerably from higher temperatures, and the potential is there for major rises in sea level, even if emissions are curtailed. During the 1990s, and up to 2011, the ice sheet was shedding around 76 billion tonnes of ice a year, enough to raise global sea level by a tiny 0.2 millimetres annually. Between 2012 and 2017, however, the rate of melting tripled to 219 billion tonnes annually, sufficient to add 0.6 millimetres to sea level.

Even at the new rate, the melting Antarctic ice contribution to sea level rise over the next 25 years would only be 1.5 centimetres. Not enough to worry too much about in its own right. If, however, the *rate* of increase is maintained over this period – in other words if the melting rate continues to triple every five years – the result would be very different. In these circumstances, the annual rise, as soon as the early 2040s, would be close to 5 centimetres a year. And this would be without any growing contribution from Greenland and from the increasing expansion of sea water as the oceans heat up further.

Although the Antarctic Ice Sheet is a continuous frozen mass, it is best treated as two: the East Antarctic Ice Sheet (EAIS) and the West Antarctic Ice Sheet (WAIS). These are separated by a mountain range known as the Transantarctic Mountains, which acts to guide ice movement. To the east of the mountains the ice flows towards the east, while on the opposing side it flows westwards. The vast majority of the ice is locked up on the east side, sufficient to raise the world's oceans by 53 metres if it all melted. The WAIS is considerably smaller and includes the Antarctic Peninsula – the rapidly warming finger of land that points northwards towards the tip of South America.

It is this western ice body that is most unstable and where most of the recent melting has occurred. It is the potential collapse of this ice sheet too, which is most worrying, in terms of its capability to raise global sea level by around 5 metres. The weight of the WAIS has forced the underlying crust downwards by as much as 1 kilometre in places, so that its base is well below sea level. As such, the

WAIS is classified as a 'marine-based' ice sheet, much of the edges of which end in giant floating ice masses known as ice shelves. As in the case of Greenland's tidewater glaciers, therefore, ice melting is being accelerated not only by rising air temperatures, but also by warmer oceans.

A critical factor in determining if and how quickly the WAIS will start to collapse is the behaviour of these floating ice masses, in particular the huge Ross Ice Shelf, which is about the size of France, and the slightly smaller Ronne–Filchner Ice Shelf. Both are now melting, and there is concern that they will eventually disintegrate. Because they are floating on the sea, and because ice takes up more space than liquid water, this would not add to sea level. The problem is, the glaciers held back by these ice shelves would be free to move considerably more rapidly seawards – two to three times as fast according to some estimates – and their break-up and melting would raise sea level significantly.

In addition to the ice shelves, the WAIS is also drained by a couple of stupendous glaciers – the Pine Island and Thwaites – and it is here that a substantial proportion of Antarctic ice loss has been focused in recent decades. Because these glaciers are not held back by ice shelves, there is nothing to stop them melting ever faster as air and sea temperatures continue to rise.

The Pine Island Glacier drains 10 per cent of the WAIS and is the fastest melting in Antarctica, on its own responsible for a quarter of all ice loss across the continent. If it melts in its entirety it will add half a metre to global sea levels, and the bad news is that it could be approaching a tipping point within the next decade or so, after which nothing we do to cut emissions will stop this.

Equally disturbing is the news from the neighbouring Thwaites glacier, a vast ice stream draining an area bigger than the UK, which is at risk of breaking up due to the effect of warm seawater attacking its base. Thwaites already accounts for 4 per cent of all global sea level rise and, if it melts completely, it will add a whopping 65 centimetres to the level of the oceans. Thwaites is also known as the Doomsday Glacier, because its demise could promote the collapse of neighbouring glaciers, eventually causing a global sea level hike of 3 metres or so. The latest research suggests that, like Pine Island, the glacier may reach a tipping point and begin to collapse within the next few years, beyond which wholesale melting is inevitable. As ever, how quickly the crossing of such tipping points translates into sea level rise is a matter for debate. The consensus is hundreds of years – at least. Others, taking a cue from the very rapid sea level hikes of the last interglacial period, think it could happen far sooner, perhaps even later this century.

Scariest of all, however, is what is happening – across the Transantarctic divide – to the vastly bigger EAIS. Until recently, the EAIS was widely regarded as a sleeping giant, pretty much immune to the ravages of global heating, but now it seems that this is not the case. In fact, six times more ice is melting in eastern Antarctica now than 40 years ago. If the end game sees both the EAIS and the WAIS beginning to collapse then – in terms of how high sea levels could climb, and how quickly – all bets are off.

STINGS IN THE TAIL 7

Despite the thousands of researchers working in the field, and the huge mass of data that has been accumulated, there are still elements of climate change science that are poorly understood or whose workings are hard to pin down. In some cases, these take the form of potential 'stings in the tail', the manifestations of which are hard to anticipate, and which could make the impacts and consequences of climate breakdown far worse. Four of the most worrying are addressed below.

The Gulf Stream falters

You may well be familiar with the Hollywood blockbuster *The Day After Tomorrow*, in which global heating triggers shutdown of the Gulf Stream and almost immediately ushers in a new ice age. Much of the film is pure hokum, but there is at least a small kernel of reality at the heart of the nonsense.

The Gulf Stream and its associated currents, together known as the Atlantic Meridional Overturning Circulation (AMOC), *is* crucial to keeping a substantial part of the northern hemisphere warmer than it would otherwise be, and its shutdown *would* bring colder conditions, but a new ice age? Categorically no!

The AMOC is best thought of as a huge conveyor belt that shifts a colossal amount of water, comparable to that of 100 Amazon rivers every second. It carries a prodigious amount of heat too – the equivalent, according to one comparison, of that generated by ten Hiroshima atomic bombs every second – and there is no doubt we would miss it if it wasn't there. All the more worrying, then, that observations suggest the smooth running of the AMOC is already in danger of being short-circuited by rising global temperatures.

In essence, the AMOC involves the northward flow of shallow, warm, salty water from the tropics and the return southward flow of deep, cold water. The transition occurs at high latitudes where sea surface waters lose heat to the atmosphere, become more dense and sink to feed the return current.

Now, however, global heating is sticking a spanner in the works by warming the sea surface at higher latitudes and driving increasing influxes of lower-density freshwater from melting polar ice. Acting together, this is curtailing the density increase that causes the northward flow to sink and feed the deep return current. The corollary is that the AMOC becomes ever more sluggish and eventually stalls.

It has been suspected for some time that the AMOC has weakened by around 15 per cent, but as this estimate was based upon monitoring over a limited time span, it wasn't possible to rule out short-term variation as the cause. Now, though, two research reports published in 2021 have shed more light on the issue and make for disturbing reading. One reveals that the AMOC is at its feeblest for more than 1,000 years, and that weakening has accelerated since the 1960s. The second really does sound the alarm bells, claiming that the AMOC has experienced 'an almost complete loss of stability over the last century' and could be approaching complete shutdown.

This goes a long way towards contradicting the most recent IPCC report, which still claims that although the AMOC is very likely to continue to slow, perhaps substantially, there is 'medium confidence' that there will not be an abrupt collapse before the end of the century. This is far from saying that it won't happen, and the latest research certainly does nothing to inspire confidence that everything is OK. My personal take is that AMOC shutdown sometime in the next 80 years or so would come as no surprise.

Big questions remain, however. Most critically, how quickly would shutdown happen? And what would be the ramifications? Answers to both questions may be provided by taking a trip back a little less than 13,000 years to a time when a short-lived cold snap – known as the Younger Dryas – punctuated our world's emergence from the last ice age. This temporary return of the cold is thought to have been triggered by the catastrophic decanting of a vast body of glacial meltwater in North America, known as Lake Agassiz, into the Arctic Ocean. The release of trillions of tonnes of icy water caused the AMOC to shut down, or at least slow dramatically, closing off the main supplies of tropical heat to the North Atlantic region, and causing the regional climate to flip.

Over a period that might have been as short as a few months, temperatures crashed as a result and near-Arctic conditions returned to Europe and parts of North America, the cooling effect then moving out across much of the planet. This is comparable to the speeded-up deep freeze that brought human civilisation to its knees in *The Day After Tomorrow*, and it does smell more of sci-fi than science. But the circumstances were not unique. As discussed earlier, the same thing happened 8,200 years ago, when an even bigger influx of glacial meltwater again acted to choke off the AMOC and knock back rising global temperatures in the 8.2 ka event. This time, the planet was warmer than in the Younger Dryas, so the effect was not as great. Nonetheless, a 400-year long cooling of the North Atlantic region followed, alongside other climatic impacts that extended across the globe. One of these was the aforementioned centuries-long drought in parts of Africa, the Middle East and Asia, which is thought to have brought people in ancient Mesopotamia together in larger groups to form the world's first towns.

But what about the future? What can we expect if the AMOC shuts down again? Modelling suggests that this would, as might be expected, delay the full extent of global heating, but this minor plus would be counteracted by a whole raft of dire ramifications. Certainly, there would be rapid and serious cooling across the North Atlantic region, driving down crop yields in Europe and eastern North America. Powerful winter storms would increase in Europe too, and sea levels would increase rapidly by up to half a metre along North America's eastern seaboard, as a result of the backing-up of the AMOC's northward flow. Further afield, the wider ramifications for global weather patterns mean that a drying out of the Amazon rainforest could be one consequence, along with disruption of the tropical monsoons, upon which billions rely for keeping crops watered.

All in all, it is a pretty bleak prospect. What keeps me awake at night is the possibility that if the AMOC shuts down, it might well remain in this state for some considerable time – possibly centuries – even if we did manage to reverse global heating.

From sink to source

Even if the AMOC continues to do its job, there may be other nasties in store, the biggest of which is undoubtedly the possibility that so-called carbon sinks, which suck up much of the carbon dioxide we pump out, will stop working. So far, luckily for us, much of the carbon dioxide generated by human activities does not remain in the atmosphere, but is absorbed by the Earth's soil, vegetation and the oceans. The trouble is, as we continue to heat up the planet, these sinks show all the signs of becoming less efficient absorbers of carbon and may even start to become carbon emitters or sources of carbon. Such a switch is an example of a positive feedback loop of global heating that has already begun in earnest.

Soils are continuing to take a real hammering from intensive farming and the widespread application of pesticides and herbicides, and in many parts of the world, including in the UK, Europe and the United States, they are effectively sterile in the absence of regular fixes of artificial fertiliser. Bearing in mind, however, that the world's soils, and the plants they support, soak up a third of carbon emissions, we continue to treat them with disdain at our peril. The fact is, soils store twice as much carbon as the atmosphere. Imagine the state we would be in if that carbon began to be released alongside human emissions. Soils also contain three times as much organic carbon as living plants and hang on to it for much longer. Plants lose their carbon when they die and quickly decompose, whereas soils can keep hold of theirs for centuries.

It was widely assumed that a hotter world would lead to soils being able to absorb more carbon, but new research shows that this assumption is completely wrong. Instead, it seems, higher temperatures are having the opposite effect. Experiments involving exposing soils, trees and other plants to levels of atmospheric carbon dioxide 50 per cent higher than they are today promoted forest growth by more than a fifth but failed to push up the amount of carbon in the soil. Instead, the additional plant growth extracted more nutrients from the soil, with the aid of symbiotic microbes in their roots, a process that actually released more carbon dioxide into the atmosphere rather than keeping it locked away in the soil. One recent study predicts that, without a rapid fall in emissions, higher temperatures, leading to an increase in microbial activity in soils, could discharge at least 55 billion tonnes of carbon by the middle of the century; near enough equal to the likely emissions of the United States over the period.

The trees and other plants that grow in the soils are also switching from carbon sinks to carbon sources due to the connivance of higher temperatures and direct human interference. Ten of the UNESCO World Heritage forest sites, for example, including the Sumatran rainforest, Yosemite National Park and Australia's Blue Mountains, have all released more carbon than they have absorbed in the past twenty years. The two biggest culprits are illegal logging and the growing number of wildfires.

One of the most deeply worrying dispatches from the climate front in 2021 revealed that the same thing was happening in the Amazon. The greatest tropical forest on the planet has played a huge part, historically, in ferreting away a significant chunk of the carbon released by humankind's thoughtless and polluting activities, but not anymore. Between 2010 and 2018, forest growth in the Amazon sucked up around 500 million tonnes of carbon dioxide annually. At the same time, however, the grubbing up of the forest on a massive scale, alongside increasingly common and severe fires, spewed out three times this amount. So, instead of being one of the world's most important carbon sinks, the Amazon rainforest is now a source, adding a billion tonnes of carbon dioxide to the atmosphere every year. With higher temperatures and drier conditions driving more wildfires, and no sign of an end to illegal logging, this is a contribution that will inevitably rise.

The precarious situation of the Amazon rainforest was also highlighted in 2021 in a major report that warned that the region was teetering on the edge of a tipping point that could have catastrophic and irreversible consequences for the world. Pointing out that one-third of the world's biggest tropical forest was already degraded or deforested and continuing to dry out, the report's authors flagged its critical importance to the planet. As we struggle to come to terms with the arrival of hothouse conditions, and given the fact that two-thirds of the world's tropical rainforest has already been destroyed or degraded, the loss of the Amazon could be the final nail driven into the coffin of our temperate climate.

As long as global heating continues, plant life as a whole will ultimately switch from carbon sink to source. This is because at higher temperatures, photosynthesis – the process whereby plants use carbon dioxide and water to make food – will begin to decline, while respiration – the means by which plants release carbon dioxide to the atmosphere – will start to accelerate. The latest research suggests that the vital carbon sink provided by land plants may be only half as effective in as little as 20–30 years, which is absolutely dire news.

So much for the land. The oceans have been doing their bit too, soaking up around one-third of all carbon emissions so far generated by human activities. This is making the upper levels of the oceans progressively more acidic as a result, threatening the survival of coral reefs and other marine life. The problem is, higher global temperatures are reducing the ease with which carbon dioxide is mixing into deeper water, so that it is becoming increasingly concentrated in a shallow surface layer. This can only absorb so much carbon, so that – at some point – it will stop taking it up, leaving more to accumulate in the atmosphere.

If and when this happens, it could switch global heating on to a trajectory that leads to a global average temperature by the end of the century 4.5°C higher than in Arkwright's day, perhaps even more. Reduced mixing of ocean water would also mean that the surface layer would get ever hotter, feeding stronger and more destructive tropical cyclones, and having a potentially calamitous impact on the world's fisheries.

The fundamental point to take away from all this is that failure to take action *now* to slash emissions may well see sink-to-source tipping points crossed in the next few decades that steers our world on to a path to society-busting hothouse conditions that we can do nothing about. Talking grandly about a net-zero world in 2050 is all well and good, but by then it may be far too late.

Methane bombs

One of the greatest natural carbon sources is locked away beneath the frozen wastes of the high Arctic, where a climate 'bomb' is about to go off that will advance global heating by decades almost overnight and have a devastating impact on the world's economy – or is it? One of the biggest, and most crucial, areas of climate breakdown debate centres on the occurrence of massive, sudden, methane 'bombs' or 'burps' exploding from thawing Arctic permafrost, and it is a debate that is far from settled.

As I mentioned earlier, methane is a hugely potent greenhouse gas, having a heating potential 86 times that of carbon dioxide. Fortunately, it hangs around in the atmosphere for much less time. Nonetheless, even over a 100-year period, it has a global heating potential 34 times greater than carbon dioxide, which means that 1 tonne of methane is as effective at heating the planet as 34 tonnes of carbon dioxide.

As the atmospheric level of methane is much smaller than that of carbon dioxide, it is measured in parts per billion (ppb). Given its potency, however, the critical measurement is the *increase* in the atmospheric concentration of the gas. In Arkwright's day, this was in the range of 600–800 ppb. Since around 1900, however – coincident with large-scale industrialisation and the intensification of agriculture – it has climbed rapidly to almost 1,900 ppb, a figure that is now rising by several parts per billion a year. A level of atmospheric methane above about 1,250 ppb contributes to excessive heating of the planet, so pushing levels back down is crucial if global heating is not to be exacerbated even further.

Annual global emissions of methane total an estimated 570 million tonnes. Around 40 per cent of this comes from natural sources like wetlands, ocean sediments and – believe it or not – methane-farting termites. The remaining 60 per cent is derived from human activities, notably in agriculture and fossil fuel exploitation, and the reduction of such emissions – as 'promised' at COP26 – could be a big factor in bringing global heating to heel. The thing is, there are also vast quantities of methane trapped in solid form – known as clathrates – in marine sediments and beneath the Arctic permafrost, which could burst to the surface as global temperatures continue to climb.

According to one estimate, a prodigious 1.4 trillion tonnes of carbon could be locked up in the form of methane and methane compounds beneath Arctic submarine permafrost. One geographic area that has attracted particular interest is the East Siberian Continental Shelf, where methane has been observed bubbling up through the frozen seabed via defrosted conduits known as taliks. Concern was raised almost a decade ago when a pair of research papers transported the whole issue into the limelight. The first proposed that as much as 50 billion tonnes of methane could be available for sudden release from the East Siberian Continental Shelf at any time, which would increase the methane content of the atmosphere twelvefold. The second suggested that a discrete methane 'burp' on this scale could advance global heating by 30 years and cost the global economy $60 trillion – a figure close to four times the US national debt.

But how realistic is this nightmare scenario? It is certainly strongly contested, and research papers published in recent years have tended to play down the methane bomb threat for a variety of reasons. These include the observation that no evidence has been found for such methane outbursts at times when temperatures were somewhat higher than they are today, notably during the previous interglacial period, the Eemian. As the saying goes, however, the absence of evidence is not evidence of absence, and some researchers suggest that it may not be possible to detect discrete methane burps – even big ones – in the geological record.

There are, however, other objections to the idea of an imminent, climate-busting methane bomb. Part of the concern is due to seriously elevated methane levels measured across the Arctic, with the sea reported to be 'boiling' with the gas in places. There is no question that thawing permafrost, both beneath the tundra and under the sea, is leading to increasing seepage of methane, but recent monitoring has revealed that most additional methane is coming from the Siberian gas fields and other fossil fuel operations.

The threat of colossal methane outbursts has not gone away, and it will continue to grow as long as we go on heating the planet. The general feeling is, however, that we can kill two birds with one stone if we focus on bringing down methane emissions arising from human activities, which would reduce the level of this potent greenhouse gas in the atmosphere *and* slow the rate of global heating. This, in turn, would act to somewhat reduce the risk of major natural outbursts of the gas.

One of the few positive outcomes of the COP26 climate conference was an agreement to cut methane emissions due to human activities by 30 per cent compared to 2020 values. One hundred and eleven countries have signed up to the so-called Global Methane Pledge, which focuses on regulatory measures requiring fossil fuel companies to check for and stop leaks of the gas from facilities. If successful, the initiative could make a fair-sized dent in the rate of global heating. If measures could be taken a bit further, so that methane emissions were cut in half by 2030, for example, the global average temperature rise could be reduced by 0.3°C in just fifteen years, which would be a very big win.

The problem, as ever in relation to global heating 'agreements', is that the signatory nations have simply *pledged* to cut methane emissions. There is no requirement to do so under international law, there are no sanctions for failing and there is no effective monitoring. Furthermore, the world's top methane emitters, including Russia, China and India, are not signed up. Nonetheless, for the first time, there is some hope that emissions of a major greenhouse gas could be reduced quickly and significantly, slowing the rate of global heating – and keeping those methane bombs at bay.

Did the Earth move for you?

Methane is not the only subterranean threat that may reveal itself in a hotter world. The idea that a changing climate can motivate the ground to shake, volcanoes to explode and tsunamis to crash onto unsuspecting coastlines seems to border on the insane. It certainly takes some getting used to, but it is, nonetheless, true.

Twenty thousand years ago, our planet was an icehouse. Temperatures were down 6°C, kilometre-thick ice sheets buried much of Europe and North America and sea levels were 130 metres lower. The following fifteen millennia saw an astonishing transformation as our planet metamorphosed into the temperate world upon which our civilisation has grown and thrived.

One of the most dynamic periods in Earth history saw

rocketing temperatures melt the great ice sheets like butter on a hot summer's day, feeding torrents of freshwater into ocean basins that rapidly filled to present levels. The removal of the enormous weight of ice at high latitudes caused the Earth's crust – which had been forced down by up to a kilometre – to bounce back, triggering massive earthquakes in Northern Europe and provoking an unprecedented volcanic outburst in Iceland. A giant submarine landslide off the coast of Norway, triggered by a quake linked to the rapid post-ice age uplift of Scandinavia, sent a tsunami crashing into the Shetland Isles and the eastern coast of the UK. Meanwhile, the enormous load exerted on the crust by soaring sea levels promoted a worldwide outburst of earthquake and volcanic activity in coastal regions.

In many ways, this post-glacial world of geological mayhem mirrors that projected to arise as a consequence of global heating. As the remaining great ice sheets of Greenland and Antarctica melt ever faster, and sea level rise accelerates, our planet's crust is once again becoming stressed and strained in response to the shifting of huge masses of ice and water. There are already warning signals that the effects of climbing global temperatures are inciting the sleeping giant to stir, and southern Alaska is the canary in the coal mine. Here, where a vertical kilometre of ice has been lost in the past 100 years, the removal of the immense load is already being reflected in more earthquake activity.

Key to understanding a dynamic response by the solid Earth, or 'geosphere', to global heating is the recognition that tiny environmental changes can have huge consequences if an earthquake fault is primed and ready to rupture or a volcano brim-full of magma and ready to blow. So sensitive are earthquake faults in such a situation that, as a seismologist colleague of mine is fond of telling people, they can be induced to rupture by the pressure of a handshake. Primed volcanoes are equally sensitive and can be compelled to erupt by something as seemingly innocuous as a heavy downpour. It is not difficult, then, to appreciate how a world of rapidly melting ice, accelerating sea levels and increasingly extreme weather can also be at greater risk of potentially deadly geological activity. The areas of most concern are high mountain ranges, coastal zones and the polar regions where ice loss is greatest.

In mountainous terrain, increasingly intense and longer heatwaves are already shrinking glaciers at an unprecedented rate and thawing out the permafrost that holds mountain faces together. As a consequence, major rock avalanches are becoming more common in the European Alps, Caucasus, Southern Alps in New Zealand and Alaska. In the decades ahead, such activity will present a growing threat to ski resorts and other high-altitude communities. In the Himalayas, earthquakes occurring on faults beneath melting glaciers have the potential to cause the breaching of glacial meltwater lakes. Dozens of these water bodies are growing across the region as rising temperatures accelerate glacier melting, presenting a huge and growing danger to villages downslope.

Coastal zones, within which the great majority of the world's megacities are located, often coincide with plate tectonic boundaries and, as a consequence, host more than their fair share of active volcanoes and earthquake faults. If they are primed and ready to go, loading of the Earth's crust as sea level climbs ever higher has the potential to set these off.

In case you find it difficult to imagine how higher sea levels could possibly trigger a volcanic eruption, Alaska's

Pavlof volcano offers the perfect example. Pavlof is a coastal volcano that prefers to erupt in the autumn and winter, the reason being that weather conditions push up regional sea level at these times of the year. The rise is temporary and amounts to little more than 15 centimetres – about the span of an outstretched hand. Nonetheless, the additional water load on one side of the volcano causes the crust beneath to bend, squeezing out any available magma like toothpaste out of a tube. In a similar manner, loading one side of an earthquake fault paralleling a coastline – California's San Andreas is the obvious exemplar – can increase tension across the fault, enabling it to rupture more easily.

When it comes to potentially the most hazardous geological response to climate breakdown, ground zero is – without a shadow of doubt – Greenland. Like Scandinavia during the last ice age, the weight of ice cover here is so colossal that, although its maximum elevation reaches more than 3 kilometres, the land surface beneath Greenland is forced down to below sea level in the centre. Rapid melting is, however, already allowing the crust to bounce back, and GPS measurements reveal that much of the North Atlantic region is already uplifting as a result. At some point, as melting continues, the load reduction will allow faults beneath the ice to rupture. When they do, they will – at a stroke – release pent-up energy accumulated over many thousands of years of dormancy, potentially triggering huge earthquakes that could exceed magnitude eight. As happened off the coast of Norway 8,000 years ago, this would be powerful enough to precipitate massive submarine sediment failures that are capable of launching devastating tsunamis across the entire North Atlantic basin. Disturbingly, researchers working in this field have suggested that a ramping up of earthquake activity beneath Greenland could become apparent within decades.

A credible response of the solid Earth to global heating constitutes a sting in the tail that has yet to impinge upon the consciousness of many people, even climate scientists. It is, however, already becoming clear that we are on track to bequeath to our children and their children not only a far hotter world, but also a more geologically fractious one.

CLIMATE WARS

8

Climate conflict today

Africa's Sahel region – the 1,000-kilometre-wide strip of semi-arid land that separates the Sahara Desert in the north from the wooded tropical grasslands to the south – stretches for almost 5,500 kilometres from the Atlantic to the Red Sea. Traditionally, the Sahel has been home to farmers and herders, following a semi-nomadic lifestyle that was best suited to chasing the rains. Now it finds itself on the front line of climate conflict.

Climate breakdown driven by global heating impacts upon conflict in two ways: by magnifying the scale or gravity of ongoing hostilities or by acting as a driver of new conflict, for example, as a consequence of mass migration or water shortages. Conflict may be between nations, political or ideological factions within nations or between those with different interests or occupations – for example, the farming and herding communities of the Sahel, who are now clashing over land and available water. Whatever form it takes, conflict inevitably leads to instability and growing civil strife, and large-scale displacement can follow, causing tension to spread into adjacent regions or countries.

Unsurprisingly, given its proximity to the Sahara, the Sahel has never been a stranger to drought. With temperatures climbing faster than the global average, however, along with reduced and more erratic rainfall, severe drought conditions are becoming more common. Major droughts prevailed in 2010 and 2012, while at other times intense rainfall has led to damaging floods. Changes to the climate have also been exacerbated by environmental degradation caused by overgrazing and clearing of vegetation. In countries like Mali, Niger, Chad, Mauritania and Burkina Faso, the loss of pastureland and the destruction of crops is driving growing hunger and increasing poverty, bringing communities into conflict and enforcing migration.

These factors are contributing to the widespread loss of livelihood, which is leaving huge numbers of vulnerable people increasingly exposed to exploitation and abuse by the Islamist terror groups active across the region. This all works towards further expanding a level of violence that is already leading to thousands of deaths every year across the Central Sahel. On top of this, almost 3 million people have been displaced, and more than 13 million left in desperate need of humanitarian assistance, including almost 1 million children with acute malnutrition.

In short, the Sahel is a tinderbox that threatens to explode into flames at any moment, launching an unending spiral of violence, mass migration and widespread famine across the region. Problems will be magnified further as climate breakdown bites ever harder, threatening to destabilise even more countries. By the end of the century, rainfall across the region could be down by a quarter, contributing – in Mali alone – to a fall in agricultural capacity of 30–40 per cent. Without sufficient food or water, and plagued by violence, it is difficult to see how anything can prevent the Sahel from developing into a string of failed and depopulating states as the world continues to heat up. In Burkina Faso, 5 per cent of the population – around a million people – were forced to leave their homes in 2020 and this is likely just the beginning of an historic exodus.

Populations are struggling with a conspiracy of climate breakdown and conflict in many other parts of the world too. Countries where violence has been endemic for a long period are the most impacted by a failing climate, because they simply don't have the capacity or leeway to adapt, leaving their coping mechanisms threadbare and ineffective. It is difficult enough to farm in a war zone without having to deal with increasingly erratic and extreme weather too. Conflict reduces resilience across the board, making it next to impossible to plan for, or to deal with, extreme weather events and their aftermath.

In Yemen, for example, where civil war has raged for seven years, two-thirds of the country – 20 million people – are in dire need of humanitarian assistance. Across Ethiopia and Somalia, a seemingly never-ending drought, flood and violence combo has damaged or destroyed already fragile livelihoods and is leading to the displacement of huge numbers of people – more than 2 million in 2019 alone. In southern Iraq, the long-term effects of war and a deteriorating climate have led to widespread environmental damage, including the drying out of wetlands and the loss of two-thirds of its 30 million palm trees, which is driving the expansion of the desert. This, in turn, is fomenting a massive exodus from the region. Loss of agricultural livelihoods in Iraq is also seen as an important factor in increased support for terrorist groups.

Afghanistan is heating up more rapidly than the global average, bringing water scarcity that is spurring farmers to focus more on growing opium poppies rather than crops that demand more water, such as almonds and wheat. Drought conditions, which sabotage livelihoods, are also forcing young people off the land and making them more amenable to recruitment by armed groups and militias.

Even where endemic violence has not – so far – been a particular issue, the impact of climate breakdown on life and livelihoods can provide fertile ground for popular discontent, which can very easily translate into violence. It has even been proposed, somewhat controversially, that the onset of severe drought conditions linked to global heating may have played a critical role in the onset of the Syrian Civil War. Although disaffection with the Assad regime was undeniably the principal cause, the worst drought in almost 1,000 years may well have seriously inflamed the situation. Drought conditions became established during 2006 and continued into 2011, when fighting started in earnest. Over the period, the country's farming sector was eviscerated. More than three-quarters of Syrian farms failed, and in parts of the country 85 per cent of livestock died, motivating the wholesale movement of 1.5 million people from the land and into the major urban centres, notably Damascus and Homs, wherein a comparable number of refugees from the Iraq War had already been shoehorned. Standards of living fell precipitously, commodities were pricier and increasingly difficult to get hold of and stress levels were high. The stage was set for grievances to be aired in public, en masse and against the government.

Future flashpoints

Even if it only acted as a magnifier of disillusionment and dissent, the Syrian situation provides a measure of just how devastating the consequences can be for a nation once climate breakdown rears its head. Over a decade later, the civil war still rages on. More than half a million lives have been lost and 12 million people have been internally displaced or driven from the country. The awful reality, however, is that this may all be insignificant in comparison with future battles over land, resources and the right to a decent life, to be fought on a hotter planet.

In truth, nowhere will be completely insulated from conflict on hothouse Earth. Violence and fighting are often no respecters of borders and even small wars have a nasty habit of spreading and punching far above their weight in relation to their destabilising effect, both regionally and on global society and economy. One estimate suggests that, for every 1°C of global heating, the risk of conflict – from social unrest and civil strife to all-out war – rises by 14 per cent. Clearly, a precise figure is not especially meaningful or useful, but such a trend would not be a surprise.

The obvious drivers of climate conflict are competition for water and fertile land and the wholesale decanting of populations from one place to another, but there are other factors too. The world's major fossil fuel exporters, including Russia and the oil-producing states of the Middle East, could, for example, suffer major economic problems as the demand for oil and gas falls during the transition to a global economy built upon renewable energy. It takes very little for failing economic circumstances to translate to social unrest. More than twenty countries depend upon fossil fuels for more than

50 per cent of export revenue, including less resilient states such as Chad, Algeria, Iraq and Nigeria, whose economies could face complete collapse if and when the fossil fuel bubble bursts.

Competition for the resources needed to be successful in a greener world also has the potential to bring nations into conflict or promote internal strife. Lithium, cobalt, manganese and graphite for the manufacturing of electric vehicle batteries will be especially sought after, alongside rare earth elements for electric motors and wind turbines and polysilicon for solar panels.

With the Arctic heating up more rapidly than almost anywhere else on the planet, the sea ice on its way out and the high latitude shipping routes opening up, greedy eyes are turning towards the region's resources. In addition to offshore oil and gas resources, which may well be left in the ground due to the cost of extraction and national carbon reduction pledges, there is also an estimated $1 trillion worth of minerals, precious metals and onshore oil and gas deposits that are already being coveted by some. Both commercial and military presences in the area are predicted to increase dramatically in the decades ahead, bringing a number of nations into strategic competition and potential conflict. It is not surprising, then, that the eight nations that claim sovereignty over lands within the Arctic Circle, including Russia, the United States, Canada, Norway and Sweden, are warning off circling non-Arctic countries, such as China, India and Japan, who are keen to share in any future spoils.

Looking at the bigger picture, the US National Intelligence Council flagged, in its 2021 assessment, eleven countries and two regions as being especially vulnerable in terms of intensifying climate impacts translating into instability and conflict. Unsurprisingly, all of the countries highlighted are majority world nations, almost half of which are in South and East Asia, including Pakistan, India and Afghanistan.

The two regions recognised as the world's most exposed to climate breakdown are Central Africa and the small island states of the Pacific. The problems of the former have already been addressed and are bad enough. Even worse, however, is the future faced by the low-lying Pacific islands, where the manifold impacts of climate breakdown are forecast to bring about economic decline and ultimately – for some at least – permanent evacuation. Another hotspot is Central America, where a combination of extreme poverty, climate breakdown and gang violence is already driving a northward exodus that brings the prospect of growing unrest and violence in Mexico and along the US border.

There is no easy way to say it, but the world of our children and their children will be a far more perilous one. As resources and habitable land diminish, nations will turn against one another in an effort to maintain or gain what they feel is their share and their right. As economies degrade, the social fabric begins to fray and mass migration becomes a global phenomenon, so the election of populist leaders promising the Earth is likely to become increasingly commonplace. At a time when the need for stable and sensible governance will never have been so critical, many countries could be led by unqualified premiers for whom posturing and sabre-rattling replace discourse and cooperation. This is nothing less than a recipe for war, and nowhere is the risk higher than between those South Asian and South-east Asian nations that depend for water and irrigation upon the great rivers that drain the Hindu Kush Himalaya mountains.

When the great rivers fail

Forget oil and gas, forget wealth, the most valuable and essential asset on the planet is water. Without it, there would be no society or economy. Indeed, without it, there would be no life at all – human or otherwise. Sure, in a hotter world, extreme rainfall events are more common and the sea encroaches steadily and remorselessly upon the land. At the same time, however, ice sheets and glaciers collapse, droughts increase, deserts expand and – most critically – rivers fail.

Earlier, I addressed the unsurvivable heatwaves that lie in wait later this century for unsuspecting populations, particularly across Asia. Without massive cuts to global greenhouse gas emissions now, three-quarters of the population of India will be exposed to 'extremely dangerous' levels of humid heat, while 400 million inhabitants of China's northern plain will be at severe risk of heat death. These are pretty grim predictions in their own right, but it doesn't end there.

Asian nations like India, China, Vietnam, Bangladesh and Pakistan all have large to very large populations, which – for India and China – hover around the 1.4 billion mark. Keeping such huge numbers of people fed and content is only possible while there is a reliable food supply, which, in turn, is critically dependent upon a trustworthy water source. Concern has been growing for some time that a failing climate will bring a more sporadic monsoon, but an even greater threat to water security is waiting in the wings. Notwithstanding the reliability of future rainy seasons, the ability of the governments of India, China and several other populous Asian nations to protect their people from hunger falls almost completely within the gift of the manifold glaciers and ice fields of the 3,500-kilometre long Hindu Kush

Himalaya mountain range. So, if and when the glaciers go, so does the water supply.

The great rivers of Asia, including the Indus, Ganges, Mekong, Yangtze and Yellow, are all fed by the ice that roofs the Hindu Kush Himalaya range. Together they supply the water to irrigate crops that feed 2 billion people across the region. It doesn't take much imagination to appreciate that if the glaciers disappear, the stage is set for a potentially cataclysmic collapse in agriculture from Afghanistan and Pakistan in the west, to Myanmar, Vietnam and China in the east.

With 15 per cent of the ice gone in the range since the 1970s, problems are already becoming apparent due to more erratic river flow, but this is only the beginning. A recent landmark report revealed that the glaciers are in such a parlous state that a business-as-usual emissions scenario will see up to two-thirds of them disappear by the end of the century. Even if, by some miracle, we managed to rapidly bring down emissions enough to keep the global average temperature rise below 1.5°C, one-third of the ice would still be gone by 2100.

Over the course of the next 30 years or so, river flow will actually ramp up as more and more meltwater cascades down from the mountains, bringing far more frequent and more intense floods at lower altitudes. Population centres will also face existential danger from catastrophic deluges arising from the breaching or overtopping of high-altitude meltwater lakes.

The real problems, however, are forecast to set in from around the 2060s onwards, when river flows will start to drop off in earnest as the glaciers fade away. Not only will this have a terrible impact on agriculture, it will also ensure that hydropower dams on the rivers can no longer function, cutting power across the region. The prospect of billions facing the double whammy of being unable to feed themselves while at the same time having insufficient power to attend to the basics of life doesn't bear thinking about. Neither does the expectation that such massive economic and societal stress is likely to lead to serious conflict.

The battle for water reserves is already feeding discontent among some of the Asian nations set to be hardest hit when the glaciers vanish. A barrage constructed by India across the Ganges River in 1975 has long caused both drought and flood problems downstream in Bangladesh, and it is increasingly blamed for elevated salinity levels and reduced water quality. Bangladesh is also not keen on India's massive water management project that seeks to divert up to one-third of the flow of the Brahmaputra, Ganges and other rivers to increasingly drought-prone regions in the south of the country.

India stores just 30 days of rainfall nationally – compared to 900 days for most developed nations – and with ground water reserves deteriorating rapidly, it is hardly surprising that the country is deeply committed to maximising its river water supply. This is one reason why the government is incensed at China's ongoing construction of three dams along the higher reaches of the Brahmaputra River, claiming it will result in irregular flow.

India is also locked in a serious dispute on its western border, this time with Pakistan, over how the waters of the Indus and neighbouring rivers are shared out. Pakistan is one of the most water-stressed countries in the world, and the Indus is its key water resource; its flood plain sustains farming that supplies 90 per cent of the country's food and accounts for 65 per cent of its employment. A water treaty has been in place with India for more than half a century, but India is now demanding a bigger share in the river's resources. Each country has plans for dam-building that the other opposes, leading to shots being fired in Kashmir. India has also issued threats about taking measures to block Pakistan's share of water from the Indus and other rivers.

As water supply concerns become ever more widespread and critical, it is easy to see how disputes could quickly escalate into conflict. Elsewhere in the world, both the Nile in East Africa and the Jordan in the Middle East have been pinpointed as possible future flashpoints, but South and East Asia remain by far the greatest areas of concern. Let's face it, as we struggle with the manifold consequences of a failing climate, the last thing we need is a hot war over water between two of the world's nuclear powers.

Exodus

On a dismal, freezing day in late November 2021, a fishing boat spotted several people in the sea off the northern coast of France. Worst fears were soon confirmed as a search and rescue operation by British and French authorities pulled the bodies of seventeen men, seven women and three children from the icy water. All were migrants seeking asylum in the UK, who had set off that morning in a flimsy inflatable completely unsuited to the dangerous and unpredictable waters of the English Channel.

More than 28,000 desperate people made the hugely risky sea journey to the UK in 2021, up threefold on the previous year, but this is just a miniscule fraction of the number of people who are on the move around the world. The situation in 2020 provides a snapshot of a catastrophe that can only get worse. The number of refugees worldwide reached the horrific total of 26 million, but this figure sheds light on only a small part of the picture. According to the Norwegian Refugee Council, many millions more people have been displaced within their own countries. Across the world, an estimated total of 55 million people were living away from their homes and livelihoods in 2020 – an extraordinary one in 143 people – 48 million displaced by conflict and violence and 7 million as a consequence of disasters, mostly linked to floods, storms or drought that we would, in an earlier time, have defined as completely natural events.

However, a different picture emerges when looking at the number of *new* displacements during the course of 2020. Reinforcing the ubiquity of extreme weather being super-charged today by climate breakdown, many people were displaced on more than one occasion, so that the true figure for displacements due to – overwhelmingly weather-related – disasters in 2020 is nearer 31 million, compared to 10 million as a result of conflict and violence.

Since 2010, it has been estimated that more than 200 million people have been forced to move, either permanently or temporarily, internally or across national borders, due to the impacts of climate breakdown. Looking ahead, the World Bank has come up with a figure of 216 million people who will be internally displaced by 2050, due to slow-onset climate breakdown impacts, such as drought. This figure, however, does not include either Europe or North America, nor does it attempt to pin down the numbers of those forced to leave their homes as a result of extreme weather, so that the reality will certainly be much worse. This view is backed up in a report by the Institute for Economics and Peace, which predicts that a mind-boggling 1.2 billion people could be displaced by 2060 as a consequence of climate breakdown, rising to 2 billion by 2100. Whatever the true figure proves to be, I think it is pretty safe to say that it will be a very big number.

There is another issue upon which the number of those permanently displaced across borders by a failing climate is contingent – both to date and in the future – and that is the meaning of the term 'refugee'. Currently, a refugee is defined in international law, in a rather long-winded way, as someone with a

> well-founded fear of being persecuted for reasons of race, religion, nationality, membership of a particular social group, or political opinion, is outside the country of his nationality and is unable to or, owing to such fear, is unwilling to avail himself of the protection of that country.

Nowhere in this definition is there a mention of climate, and this is a problem because it means that those forced to leave their country as a consequence of climate breakdown cannot, legally, be called refugees. Instead, they must be classed not as 'climate refugees', but as 'climate migrants'. Refugee status, however, confers certain rights that migrants miss out on, leaving them more vulnerable and providing countries with a ready-made excuse to exclude them.

The problem could be alleviated by coming up with a legal definition for climate migrants, but this is far from straightforward. Would such a migrant, for example, be someone forced to leave their country as a direct consequence of climate breakdown, or someone whose livelihood was destroyed as a result of a chain of events, within which a failing climate was just one link? It may take a while to resolve this issue, but it must be resolved if the many millions of people forced from their homes by climate breakdown are to enjoy the help and protection they deserve.

Whatever we call those whose way of life is destroyed – either directly or indirectly – by a broken climate, there is no doubt that en masse they are slated to be a hugely destabilising factor right across the planet, feeding feelings of otherness and animosity in indigenous populations and provoking social unrest, civil strife and, in the worst cases, all-out war. And don't kid yourself that this will be confined to majority world countries alone. Even as climate migrants from African nations arrive in increasing numbers in Southern Europe, so future blistering heat, widespread drought and crop failures will likely drive a permanent movement of people north from Greece, southern Italy and Spain.

A 2-metre sea level rise, which is perfectly possible within 80 years, would see one-third of the population of Florida's Miami-Dade County – not far short of 1 million people – having to up sticks and move away. In the UK, a comparable rise could see the inhabitants of Lincolnshire communities such as Boston and Spalding seeking new homes and new lives. By the second half of the century, 40°C+ summer temperatures in southern and central parts of the UK could well be driving people out of their appallingly insulated homes and towards cooler parts of the country.

Pretty much everywhere in the world, the arrival, in large numbers, of climate migrants is certain to stoke hatred and grow support for anti-migration populist politicians, parties and regimes. It is a recipe for a human catastrophe that could well dwarf the scale of direct physical impacts due to a collapsing climate and potentially tear the social fabric of many nations apart.

HEALTH AND WELL-BEING ON AN OVERHEATED PLANET 9

Feeling the burn

Somehow, it seems intuitively right to link more warmth to better health, but a hotter world is not a healthier one, quite the opposite in fact. Of the 5 million or so people who died every year over the past two decades as a consequence of extreme temperatures, more actually died of excessive cold than overwhelming heat, but this is changing fast. Between 1991 and 2018, more than one-third of heat-related deaths across the planet were the result of global heating, rising to more than 50 per cent in places like Kuwait, Iran, the Philippines and Central America. As might be expected, the elderly are especially vulnerable, and the annual global number of extreme heat-related deaths among the over 65s has risen by more than half in the two decades to 2018, when the figure was a shade under 300,000.

Looking ahead, the death toll due to cold will fall – although a wildly swinging jet stream will still bring episodes of freezing weather to temperate latitudes – but this will be more than offset by massively increased mortality rates due to extreme heat. In future decades, the old, infirm and very young, in particular those who don't have access to air conditioning, will suffer increasingly from rising temperatures and humidity. The latest research reveals that not even the unborn are immune. Higher temperatures are not only implicated in greater numbers of premature births, but also cause early weight gain in babies that increases the risk of obesity in later life. Another study shows that smoke from wildfires doubles the risk of severe birth defects.

Within twenty years, ferocious heatwaves, comparable with the worst experienced to date, will affect nearly 4 billion people every year – more than half the current population of the planet. In 2019, an estimated 302 billion hours of labour were lost worldwide as a consequence of increasing temperatures, up more than 50 per cent on two decades earlier, and this figure is expected to climb rapidly, severely impacting productivity and national economies. In its 'Climate change risk assessment 2021', the prestigious UK think tank Chatham House came up with the horrifying forecast that within ten years – without immediate and drastic cuts to emissions – heat stress could prevent more than 400 million people working outside. At the same time, every year an estimated 10 million people could be exposed to extreme heat that exceeds the survivability threshold.

The reality, then, is clear. We are facing a global health catastrophe. In its landmark report published in 2009, the UCL–Lancet Commission on Managing the Health Effects of Climate Change, of which I was a member, labelled climate change the biggest health threat of the century. Prior to the 2001 COP26 climate conference, 200 health journals published a joint editorial calling for world leaders to take emergency action on climate change in order to protect health and well-being. The editorial pointed a finger at world leaders, announcing that the greatest threat to global public health was their continued failure to keep the global average temperature rise below 1.5°C and to restore nature. Inexcusably, the call seems to have fallen on deaf ears.

Heatstroke will, of course, be a major killer, but a hotter climate will also bring other challenges to health and well-being. For every 1°C increase in the global average temperature, the death rate due to respiratory illness rises by around 10 per cent. Cataracts will also become increasingly common, while cases of malignant melanoma are forecast to be up 50 per cent by 2040. There are other less obvious threats too. For example, an estimated 33,000 people are now losing their lives annually to pollution associated with the ever-growing number and scale of wildfires, the risks being especially strong among firefighters.

Those working in the open air for extended periods, notably in the tropics and equatorial zone, will be most at risk. The appallingly treated migrant labourers in Qatar, for example, are often required to work in baking heat for eight hours at a time, relieved by just a 30-minute break, and with little protection from the heat and humidity. It is hardly surprising that hundreds drop dead every year. Among the workers in the sugar plantations of Nicaragua, kidney disease has been endemic for generations, and it was only recently recognised as being a consequence of heat stress and insufficient consumption of water.

Neither are those in cities off the hook. As touched upon earlier, the heat island effect means that large built-up areas are typically several degrees hotter than the surrounding countryside. Workers in urban factories that lack air conditioning will become increasingly at risk of heatstroke and related conditions as the world gets hotter, and those who simply stay at home will not fare much better. Urban housing, in both developed nations and the majority world, is simply not designed to handle the levels of temperature and humidity predicted for the middle of the century and beyond. In a survey of the administrators of 800 of the world's cities, almost half said they had no plans in place for adapting to the impacts of global heating, many pointing out that they simply did not have the funds to do so.

Then there are the mental health issues, which are especially prevalent among those caught up in the trauma of extreme weather disasters, those who have lost livelihoods or those forced out of their homes and on to the road. A study published a few years back revealed that climate breakdown was a factor in the suicides of nearly 60,000 Indian agricultural workers over the past 30 years. Highlighting the desperately precarious existence of South Asian farmers, the authors of the study showed that just a 1°C temperature rise on an ordinary day in the growing season resulted in 67 more suicides and a 5°C increase in 335 additional self-inflicted deaths. Conversely, just an extra centimetre of rainfall a year was linked to a 7 per cent fall in suicides. Since 1995, more than 300,000 Indian farmers and farm labourers have killed themselves, and without help – for example, insurance schemes to protect against harvest failures – this is a figure that is certain to grow as temperatures climb and the rains become more erratic.

There are also direct links between global heating and mental health. A 2021 research paper has, for example, linked episodes of intense humidity with higher suicide levels, especially among women and young people. As an example, those struggling with mental illness may be affected more by sleep deprivation arising from high humidity levels at night. People on anti-depressants may also feel increasingly uncomfortable due to these drugs sometimes interfering with the body's ability – under conditions of higher humidity – to regulate its temperature.

More generally, and not at all surprisingly, anxiety about the future is affecting increasing numbers of young people in countries right across the planet. The impact of climate breakdown on mental health is still going largely unseen and unrecorded, but it has the potential to become a huge issue later in the century. Today, mental health conditions affect 1 billion people worldwide and cost the global economy around $1 trillion a year. In the future, it may well prove to be one of the biggest of the many deleterious impacts of climate breakdown.

Hungry world

Every night, 800 million people go to sleep hungry. Most are in majority world countries, but daily hunger is also rising in developed nations with huge and growing wealth disparities, such as the UK and the United States. Our world's population is not rising as fast as it was. Nonetheless, it is predicted to peak at between 9 and 10 billion by the middle of the century, mostly driven by growth in majority world countries in Africa and Asia. This is the very worst news, because at the same time a medley of higher temperatures, water stress and extreme weather is forecast to progressively reduce yields of staple crops. The shortages and price hikes caused by the unprecedented summer of 2021 are forerunners of what is to come, but provide little idea of the scale of future food production problems, which is likely to be devastating.

In its 2021 climate change risk assessment, Chatham House forecast that by 2050, when an increased global population and growing demand will require 50 per cent as much food again as today, agricultural yields could be down by almost one-third. If realised, this nightmare scenario would mean nothing less than widespread and unprecedented famine and the societal breakdown and conflict that would inevitably accompany it.

And we can't say we haven't been warned. In recent years, regional harvest failures have become far more common, and this trend is increasing. In two successive years – 2006 and 2007 – the Australian wheat crop was down by half due to severe drought. Abnormally high temperatures and drought conditions in 2018 played havoc with the agriculture sector across much of Europe, slashing crop yields and driving many farmers into bankruptcy. Extreme weather in 2020 resulted in the UK wheat crop falling by 40 per cent, and the following year devastated French vineyards, slashing wine production by a third at a cost to producers of €2 billion. According to the UN World Food Programme, the number of food emergencies has climbed from around fifteen a year in the 1980s to well over 30 a year now.

The evidence that global farming is already beginning to take a beating from climate breakdown is unequivocal. In fact, since 1960, agricultural productivity has fallen by one-fifth compared to what it would have been in the absence of global heating. This trend will continue as the planet heats up and the area of land available for food production progressively reduces due to excessive temperatures, drought, sea level rise and desertification.

According to the latest forecasts, by 2100, the world's deserts will have increased in size by 4 million square kilometres – an area larger than India. In less than two decades' time, by 2040, almost one-third of global cropland will be affected by severe drought – sufficient to halve yields – in any single year, up three times on the historic average. The middle of the century will see more than one-third of the land on which rice and wheat are grown exposed to yield-reducing heat every year, this figure rising to almost two-thirds in parts of South Asia. At the same time, the breadbasket areas of both the United States and Russia are forecast to face extreme drought across 40 per cent or more of their growing areas every single year.

Harvest failure can be devastating, both locally and regionally, and this is almost always the case in the majority world. In the future, there is an increasing likelihood of climate breakdown bringing so-called multiple breadbasket failures, which involve the simultaneous loss of a significant proportion of the global harvest of one or more staple crops. According to a recent report by the consultancy firm McKinsey, the chances of at least one global harvest failure of more than 15 per cent in any decade is set to climb from 10 per cent today to 34 per cent by the middle of the century. Such an event could more than double prices, which would have an appalling impact on the world's poor and likely foster widespread social unrest and civil strife. In a similar vein, Chatham House flags the danger of a simultaneous major yield fall (by more than 10 per cent) in the four main maize-growing countries (the United States, China, Brazil and Argentina) being just short of 50 per cent by the 2040s.

A snapshot of the future can be captured by looking back at the global food price crisis of 2007–2008, when the cost of rice climbed by more than 200 per cent and the price of wheat and soy beans more than doubled. The cause is still widely debated, and more than a dozen factors have been linked to the development of the crisis, including regional falls in production, financial speculation, declining world food stockpiles and the impact of extreme weather, especially in Australia. Whatever the cause, the ramifications were huge and widespread. Many nations growing staple crops banned exports to maintain a domestic supply. Meanwhile, developing countries with weak farming systems saw their food import bills climb by a quarter.

Resulting price hikes making basic foodstuffs unaffordable launched a wave of social unrest across the majority world, in places as far apart as Tunisia, Egypt, Cameroon, Bangladesh and Haiti. The biggest protests took place in North Africa, especially in Tunisia where – exacerbated by other factors – they culminated in a revolution that saw the toppling of the dictator-led government. This, in turn, launched a domino effect of popular uprising across the region, known as the Arab Spring, providing testimony to the critical importance of affordable food to the maintenance of social stability. As the American writer and journalist Alfred Henry Lewis pointed out in 1906: 'There are only nine meals between mankind and anarchy.' Some would say even fewer.

All this leads to a horrific vision of a world, later this century, in which growing famine and food insecurity will ensure that billions are forced to vie for the calories they need to survive, and many will lose the struggle. One study suggests that, unless emissions are rapidly reduced, fully half of the world's population will face severe food shortages by the end of the century. Across the majority world, this would mean massively increased mortality rates due to starvation and malnutrition, compounded by violence, abuse and exploitation. Even those in developed countries would suffer, with scarcities of basic commodities increasingly common and millions struggling to feed their families as food prices hit astronomical levels.

March of the mosquito

Unfortunately for us, a warmer, more humid world is just heavenly for mosquitoes and other insects that carry deadly diseases. As temperatures continue to rise, the stage is set for decades of progress to control the transmission of insect-borne diseases to be undone, leading to a hike in all sorts of unpleasant illnesses. This will come as a big shock to healthcare systems in many countries, as – according to the World Health Organization – less than half have any sort of national health and climate change strategy in place.

Diseases carried by insects are usually referred to as 'vector-borne'. In other words, they are carried by another organism (vector) instead of being communicated directly from person to person or picked up from dirty water or rotten food. They occur in various forms, including parasites, viruses and bacteria, and cause a wide range of illnesses, such as malaria, dengue fever, yellow fever, African sleeping sickness and zika. Even in today's world, such diseases take a huge toll, causing more than 700,000 deaths a year, of which over half are due to malaria. Every year, there are well over 200 million cases of the disease, alongside at least 100 million (and possibly up to 400 million) estimated instances of dengue fever infection.

The incidence of some vector-borne diseases is already on the rise. A number of countries, including Venezuela, Sudan and Eritrea, have seen a resurgence of malaria, while the number of *reported* cases of dengue fever has shot up from half a million to more than 5 million in just two decades. Looking ahead, the picture painted is pretty grim for areas where malaria, dengue fever and other diseases are endemic. This is because not only do hotter conditions encourage pathogens to replicate and mature more rapidly within the host insects, but the reproduction rates of the hosts themselves increases, leading to a higher density of insects in a particular area. With the bite frequency rising in step with temperature too, the result will inevitably be a significant jump in the incidence of insect-borne diseases.

Global heating will also make the situation worse through extending the range of many disease-carrying insects. Already, ticks, which host Lyme disease and other nasty maladies, are extending their range in Europe and North America, moving further north and to higher altitudes as temperatures climb. Malarial mosquitoes are also heading uphill, increasing the risk of malaria to inhabitants of highland areas in both Africa and South America, especially during hotter years. Later this century, in the absence of major cuts to emissions, a huge expansion of range is predicted for mosquitoes carrying malaria and dengue fever, which could have dire consequences for the health systems of countries unprepared for the diseases, and for the huge numbers of people previously unaffected, who – as a result – have no accumulated resistance.

Insect-borne diseases may also be transmitted by means of international travel. For example, while malaria is no longer endemic in the United States, mosquitoes are common. As a consequence, local outbreaks of malaria have arisen due to indigenous mosquitoes feeding on infected people, who picked up the disease abroad, and then biting others. And then there is so-called airport malaria, carried by infected mosquitoes on aircraft or in luggage and spread to the local populace after landing. As global temperatures continue to climb, such imported mosquitoes will be able to survive and even thrive across more of the Earth's surface, bringing malaria and other mosquito-borne illnesses back to countries from which they had been eliminated. Malaria was only eradicated from the UK after the Second World War, and from Italy as recently as 1970. Mosquitoes remain abundant in both countries and summer temperatures are becoming increasingly conducive to sustaining potential malaria outbreaks.

Provided the conditions are right, the speed with which insect-borne diseases can become established in a country is pretty scary. In 1999, isolated cases of West Nile virus were detected in New York. The illness is a tropical disease caused by a virus which is carried by mosquitoes, and it can also lead to meningitis and encephalitis. The source of the outbreak remains unknown, but there is a strong possibility that one or more infected mosquitoes arrived in airline or shipping cargo. Whatever the source, the initial 1999 outbreak may have infected more than 8,000 people. By 2004, West Nile virus, which was previously unknown in North America, had spread to all 48 contiguous states, and up until now, there have been an estimated 7 million cases. The disease is now well and truly endemic.

West Nile virus is favoured by higher temperatures and drought conditions, so a hotter world suits it perfectly. It is already endemic in Europe, and its occurrence is certain to become more common. Dengue fever is now endemic in parts of Europe too, notably in the eastern Mediterranean, while local cases are now reported across the continent every year. Another mosquito-borne nasty, chikungunya virus – which causes severe joint pain – is also infiltrating Europe, with outbreaks in France and Italy. Malaria outbreaks have been occurring in Greece since 2009, and it is highly likely that the disease will spread widely across Southern Europe in coming decades, and perhaps to central and northern parts of Europe too, including southern and eastern parts of the UK.

There really is no doubt that insect-borne diseases are going to be a major threat to the inhabitants of a hotter planet, both across the majority world and in developed countries. A 2021 study by the London School of Hygiene and Tropical Medicine came up with some especially disturbing predictions. For a business-as-usual scenario, involving unabated emissions, the authors of the study found that rising global temperatures would, by 2080, lead to the transmission season for malaria increasing by a month, while that for dengue fever would be four months longer. Given the expected rise in global population by this time, a staggering 8 billion people – close to 90 per cent of the world's inhabitants – could, as a result, be at risk from one or other of the two diseases.

Some hope may be gleaned from the fact that there are now reasonably effective vaccines for both malaria and dengue fever, but large-scale vaccination campaigns will become increasingly difficult to fund and institute against a future hothouse backdrop of growing poverty, mass migration, famine, burgeoning social unrest and conflict. And there is the prospect, too, of new insect-borne diseases, against which we have no protection. West Nile virus and zika both originated in other species before mosquitoes inflicted them upon humans. It is chilling to consider what other nasties are being incubated in the fauna inhabiting the jungles and swamps of a hotter and clammier world.

Dirty water, rotten food

One and a half billion people do not have access to drinkable water – almost one in five people on Earth. On a hothouse planet, more and more people will find it increasingly difficult to slake their thirst safely as water becomes an economic and political football that is fought over, hoarded and devoured by agricultural and industrial sectors struggling to keep going.

While rivers are reduced to trickles and aquifers sucked dry to support farming, even lakes are likely to become off-limits as a supply of potable water. Warmer water is able to hold less oxygen, so that standing bodies of water are already being slowly suffocated, promoting the growth of 'blooms' of toxic algae that make the water dangerous, even deadly, to consume.

As growing numbers of people, mostly – but not exclusively – in majority world nations, are forced to survive on contaminated water, cholera will become more rampant. *Vibrio cholerae*, the virus that causes cholera, is a naturally occurring bacterium found in rivers and coastal waters. It is highly contagious and is typically picked up from contaminated water or food. Cholera has plagued human civilisation for almost as long as it has existed, and – in many parts of the world – still does.

Historically, the disease has been an implacable killer, taking 1 million Russian lives over the course of just five years in the mid-19th century, and causing 8 million deaths in India between 1900 and 1920. Today, it infects up to 5 million people a year, leading to as many as 143,000 deaths, mostly of young children. In 2010, an outbreak of an especially virulent cholera strain in Haiti – capable of killing within a couple of hours of infection – took 9,000 lives. An ongoing epidemic in war-torn Yemen, driven by failing water supplies and a collapsed health system, has persisted since 2016, resulting in the infection of more than 2.5 million people and causing 4,000 deaths.

The World Health Organization has vowed to end human-to-human transmission of cholera – for example, due to a contaminated person handling food – by 2030. This is a fine ambition but set against a background of increasing social and economic instability and falling availability of clean water for many, it is hard to see how this could be accomplished. On the contrary, it is likely that more cholera epidemics will arise in the decade ahead and beyond. Cholera thrives in hot and damp conditions, so it is perfectly suited to an overheated, wetter world. Outbreaks tend to die down during cooler winter months, but longer and hotter summers and warmer oceans and rivers will extend the period over which outbreaks can be sustained. Extreme weather will play a key role too, driving disasters during which health systems struggle and drinking water becomes contaminated with the cholera bacterium.

Other dangerous vibrio bacteria too are expected to thrive as the heat builds that can inflict gastroenteritis, severe wound infections and sepsis upon those they infect. Some of the most dangerous live in coastal waters and a warmer ocean is fast expanding their range, even in places as far north as the Baltic and North Atlantic. In future, swimming in the sea, it appears, will bring a new and growing threat for many.

Speaking of stomach bugs, thinking about my last encounter with the salmonella bacterium still sends a shiver down the spine, even though it happened a good 30 years ago. Symptoms came on within hours of scoffing some dodgy coleslaw, and I spent the next week commuting hurriedly between bed and bathroom. Illogical I know, but I have never touched coleslaw since. In the circumstances, I was somewhat disquieted to discover that this water-borne bacterium is set to thrive on a hotter planet. The bug is also food-borne and forecast to cause serious problems as the heat continues to build, especially for the billions who have no access to refrigeration. I was astonished too at the impact this bacterium already has, with more than 1 billion cases of salmonella poisoning worldwide every year, leading to as many as 3 million deaths. Some strains of salmonella can also cause typhoid, of which an estimated 200,000 people die every year.

In the UK, every 1°C rise in temperature is forecast to lead to a 12 per cent increase in cases of salmonella poisoning. There also seems to be a hook-up with extreme weather, one study demonstrating a clear link between increasing salmonella cases and heatwaves or episodes of extreme rainfall. This is because such meteorological conditions both accelerate the rate at which salmonella bacteria breed and increase the chances of contamination of water and food crops.

As far as I know, I haven't yet succumbed to campylobacter poisoning, but maybe that time will come because this is another pathogen carried by contaminated food and water that is slated to prosper in a hotter world. Every year, the bacterium causes more than 800 million cases of bacterial gastroenteritis worldwide and is the most common cause of food poisoning in Europe, where cases have been increasing since 2008. One recent study has suggested that cases will double in the next 60 years or so in Northern Europe, and this trend is likely to be mirrored elsewhere. Looking for somewhere to place the blame, a piece of Canadian research has pointed a finger at the common house fly as a possible cause of a future rise in campylobacter cases. Warmer weather, so the theory goes, will cause flies to become more sprightly. As such, they will visit more food items than in cooler weather, leading to their mucky feet spreading the bacterium far and wide.

Climate breakdown can increase disease levels in more indirect ways too. Tuberculosis transmission, for example, is expected to become more prevalent as refugees, migrants and people displaced by extreme weather are forced to live cheek by jowl in camps and shanty towns. And dangerous micro-organisms will not be the only problem as the heat ramps up. Warmer and wetter winters and springs in temperate regions are predicted to drive a rise in rat populations, while heatwaves could drive thirsty rats indoors seeking water. Together, this will result in greater interaction with humans, so increasing chances of spreading all sorts of unpleasant diseases, including hantaviruses and plague.

It is disturbing, to say the least, to discover that climate fluctuations appear to have played a role in a number of plague pandemics – including the Black Death in the 14th century – that had their origins in Central Asia. Looking ahead, just a 1°C temperature rise across this region during spring could potentially hike the prevalence of the plague bacterium, *Yersinia pestis*, in black rats by 50 per cent. As in the past, future plague outbreaks in Asia would have the potential to spread further afield, including westwards to Europe and around the world.

THE BIG QUESTIONS 10

How bad could things get?

In all honesty, we might as well ask: 'How long is a piece of string?' So complex and intertwined are the interactions and relationships between the climate, the natural world and human society and economy that – despite meticulous and comprehensive modelling – the nature, scale and breadth of climate breakdown impacts later in the century, and beyond, can only be guessed at. In addition, how much we reduce greenhouse gas emissions, and how quickly, will play a huge role in determining where we eventually end up. Because it is now going to be practically impossible to keep this side of the 1.5°C dangerous climate change guardrail, what we can be certain of is that climate breakdown will be all-pervasive. Insidiously worming its way into every corner of lives and livelihoods, no one, anywhere – not even the tech billionaires in their guarded redoubts – will be immune.

In Arkwright's day, the average temperature of our world was 13.7°C (56.7°F). This rose slightly over the course of the 20th century, for which the average temperature was 13.9°C (57°F). Now it is close to 14.9°C (58.8°F) and climbing fast. If we leave it up to the fossil fuel corporations and their supporters and apologists it would just keep on going. Completely in thrall to the profit motive, and left to their own devices, they would have no compunction about hoovering up oil, gas and coal until not a dreg was left. The consequences would be calamitous, and that's putting it mildly.

In a seminal research paper published in 2013, aforementioned ex-NASA climate scientist James Hansen and his colleagues figured out what the outcome would be if we burnt all – or at least most – of the planet's known fossil fuel reserves, and their conclusion is the stuff of nightmares. Hansen and his co-authors showed that, ultimately, most of the planet would become uninhabitable, largely due to the simple fact that, over much of the Earth's surface, it would simply be too hot for humans to survive. On average, the planet would heat up by 16°C (61°F), bringing its average temperature to more than 30°C (86°F). Over land, the average rise would be 20°C (68°F), and at the poles a staggering 30°C. Such conditions would be impossible to adapt to and would not only destroy human civilisation but reduce the global population to a tiny fraction of what it is today.

Humans have done some pretty dumb things in the past, both as individuals and collectively. Nonetheless, it is hard to imagine that we would sign up to self-imposed genocide simply to put a bit more cash in the pockets of fossil fuel company CEOs. This is small consolation, however, for future generations, who will still be forced to face a very different, and far more dangerous, world than the one we have been lucky enough to grow up, or grow old, on. The reality is that even if the short-term goals set out at COP26 to bring emissions down are met, the global average temperature rise will still be at least 2.4°C and quite possibly higher. When every tenth of a degree increase brings further extreme weather, faster melting ice sheets and more rapidly rising oceans, this is a huge overshoot above and beyond the 1.5°C dangerous climate change guardrail.

How rapidly the global average temperature will climb depends upon what we do next. If business as usual continues, we could barrel through the 1.5°C guardrail in the next six to nine years and the 2°C mark in as little as twenty years. As I mentioned earlier, worse news is that, according to a research paper published in *Nature* in 2021, whatever action we take to cut emissions we will not be able to stop a temperature rise in excess of 2°C, and probably as high as 2.3°C.

It is not, however, all bad news. True enough, if we carry on the way we are, a 2°C hike could be upon us within a couple of decades. But if, instead, we act to slash emissions now, this rise could be put off into the next century or even beyond. While still far from ideal, this would at least provide much-needed breathing space during which to adapt to a hotter world, and to take measures – such as wholesale tree planting – to bring carbon levels down, so that we might not get to 2°C at all. At this point, however, it is worth reminding ourselves that this level of temperature rise would make our world hotter than during the previous interglacial period (the Eemian), when the sea level was several metres higher, so sticking to a 2°C rise, even should it be postponed, should not be regarded as a victory.

A further worry is that if the global average temperature keeps climbing during the rest of the century and beyond, however slowly, opportunities continue to exist for tipping points to be crossed that cannot be uncrossed, even if and when the temperature begins to fall again. Much higher sea levels in the Eemian, for example, suggest that there must have been substantial collapse of the West Antarctic Ice Sheet. Analysis of the Eemian climate also points to the failure of the Atlantic Meridional Overturning Circulation at some point. As touched upon earlier, this is already unstable in our world, so a breakdown would hardly be a surprise. The consequences, however, would add yet another layer of unwanted and pernicious climate mayhem.

Taking a broader perspective, tipping points and positive feedback loops are the real flies in the ointment when trying to pin down how bad things are going to get. From ice sheet collapse and AMOC shutdown to methane bombs and failing carbon sinks, these are – in a very real sense – the equivalents of former US Defence Secretary Donald Rumsfeld's 'known unknowns'. Rumsfeld's convoluted terminology was intended as an explainer for the absence of evidence that Iraq had weapons of mass destruction (it hadn't). Nonetheless, it is also fitting as an accurate descriptor of the state of current knowledge and understanding about how some parts of the climate system operate.

Can the planet hackers come to the rescue?

In my eco-thriller *Skyseed*, a clandestine attempt to engineer a way out of the climate emergency goes horribly wrong. Fortunately for us, the self-replicating, carbon-munching nanobots that plunge our world into chaos only exist – as far as I am aware – in my febrile imagination. Still, the message the book carries is still both urgent and timely: deliberately mess with the climate and it *will* come back and bite you.

This message hasn't, however, hit home among those increasingly keen on tinkering with the climate system to try to bring global heating to heel. Technological schemes that involve intentional, wholesale interference with the climate are subsumed within the broad heading of 'geoengineering', and there are a growing number of them. In the simplest terms, proposals can be grouped together into those that look to cool the planet by blocking incoming solar radiation and those that seek to do so by actively scavenging carbon from the atmosphere. Personally, I like to divide them up differently, into pie-in-the-sky and pragmatic.

Proposed geoengineering methodologies either carry huge risks, are immensely expensive or are environmentally damaging at scale. At the top of my 'pie-in-the-sky' file is a proposal called SCoPEx (Stratospheric Controlled Perturbation Experiment), which aims to cool the planet by simulating a large volcanic eruption. Developed by a group at Harvard University and backed by Bill Gates and other mega-rich donors, the reasoning behind the project is quite straightforward. Large volcanic blasts cool the climate, ergo, if such an event can be faked, maybe we can artificially do the same.

Volcanic cooling is a consequence not of the often-huge quantities of ash ejected into the atmosphere by big eruptions, but of the accompanying sulphur gases. In sufficient quantities, sulphur dioxide blasted into the stratosphere forms an aerosol mist of sulphuric acid droplets capable of forming a veil extensive enough – following the largest eruptions – to shroud the planet, or a substantial part thereof. This sulphurous screen is especially effective at reflecting incoming solar radiation back into space, so cooling the troposphere (the lower atmosphere) and surface of the planet.

The SCoPEx crowd, if they get their way, plan to simulate this effect by spraying chemicals from high-altitude balloons or fleets of aircraft, which – the team claims – could cut heating by 1.5°C at the cost of $10 billion a year, or even less. But this is only if the chemicals in the stratosphere continue to be replenished year after year after year. They insist – of course – that modelling shows the whole thing to be perfectly safe, and free of any nasty side effects. Given the enormous variability inherent in any modelling and the extreme complexity of the climate, my view is that this is nothing more than scientistic hubris. The fact is, the whole idea is plagued by uncertainty and fraught with danger, so much so that one eminent atmospheric scientist has listed fully 27 reasons why this form of geoengineering is a very bad idea. Reinforcing such concern, more than 300 experts – including me – have signed a petition calling for the UN to institute a non-use agreement on solar geoengineering.

As a volcanologist, I am very familiar with past volcanic blasts and their ability to impact seriously upon the climate. In 1783, the Laki eruption in Iceland spewed out around 120 million tonnes of sulphur dioxide, causing temperatures to plunge across Europe and North America the following year. Further afield, the African and Indian monsoons were weakened, while rainfall was reduced across the Sahel, resulting in the greatly reduced flow of the Nile. Even more impressive was the impact of the colossal 1815 eruption of Tambora in Indonesia, which drove severe northern hemisphere cooling that led to the 'Year Without a Summer' in 1816, devastating harvests and provoking the last great subsistence crisis in the Western world. The message from the past is clear: the effects of volcanic clouds, whether natural or artificial, are dangerous and unpredictable, and should be avoided at all costs.

Also in my pie-in-the-sky file is an equally wild and wacky proposal to 'refreeze' the poles by spraying seawater into the atmosphere above to create brighter clouds able to reflect more sunlight back into space. Another advocates spraying chemicals at high altitude to dissipate wispy cirrus clouds, thereby allowing more heat to escape into space. There are plenty of other schemes in a similar vein, but there is no space to address them here. What they all have in common, however, is that they are effectively large-scale experiments designed to mess with the climate system so as to sort out the mess we have already made. As such, they are hugely risky and carry the very real potential to make matters worse. All of these schemes should stay where they are today, firmly confined to the drawing board.

In my pragmatic file are schemes that carry less risk, but which nonetheless bring their own baggage. One reason-able sounding idea, called 'enhanced weathering', involves spreading vast quantities of crushed basalt rock on the world's farmland. Not only could the rock dust improve the soil, but it would also soak up carbon from the atmosphere as it weathers and breaks down. The problem is, undertaken at a scale big enough to make a worthwhile dent in carbon levels, the scheme would be hugely environmentally dam-aging and require massive amounts of water and energy in the production process, to say nothing of the transport and health issues.

Comparable schemes involve the construction of industrial plants to actively suck carbon dioxide out of the atmosphere and turn it into rock. One opened in Iceland in

2021, capable of extracting 4,000 tonnes a year. This means we would *only* need 10 million of these plants across the planet to cancel out a single year's carbon emissions. Clearly, the cost, upheaval and environmental damage will mean that this just ain't going to happen.

Perhaps the most ludicrous carbon extraction proposal is a plan to construct 10 million artificial 'trees' designed to pull carbon dioxide out of the atmosphere by means of chemical reactions. This at a time when we are hacking down the same number of *real* trees every five or six hours. The irony and absurdity make you want to weep.

Most mundane of all carbon-extracting schemes is carbon capture and storage (CCS). This involves intercepting and liquefying the carbon dioxide released at power plants or industrial installations and storing it underground, for example, in oil and gas reservoirs that have been sucked dry. At face value it seems a perfectly reasonable plan, but the bottom line is that the CCS process is very energy intensive, costly, technically challenging and carries significant risks, for example, of future leakage. Not only this, but fossil fuel companies – who are big fans of CCS – would not only make money out of developing the necessary infrastructure, but also hijack the process to squeeze out the last dribbles from senescent oil fields. In any case, the tiny amount of ongoing CCS would need to be scaled up nearly 60 times by 2030 to make any sort of impact on emissions, and this seems highly unlikely.

On top of all the aforementioned issues and risks, geo-engineering also flies in the face of the human and legal rights of every individual on Earth, who would have no say in any decision to hack the planet. It is also a distraction that seeks only to address the symptoms of global heating rather than the cause, and which detracts from efforts to bring down emissions as fast as possible. In fact, promoting the belief that there is a technological backstop we can resort to if measures to cut emissions fail makes it all the more likely that such measures *will* fail. If world leaders think bringing down emissions by conventional means is not the last line of defence, then it is inevitable that their hearts and minds will be distracted from the urgency of the job in hand by the possibility that a techie fix is there to fall back on. The consequences of such a change of focus would be nothing less than calamitous.

How can we stop a bad situation becoming worse?

A fusion of ignorance, poor governance and obfuscation and lies by climate deniers has ensured that we have sleepwalked to within less than half a degree of the 1.5°C dangerous climate change guardrail. Soon, barring some sort of miracle, we will crash through it. Today, global heating and climate breakdown sit at the top of the political agenda, thanks in no small part to inspirational individuals like Greta Thunberg, initiatives such as Fridays for Future and grass-roots activist organisations like Extinction Rebellion and Insulate Britain. While awareness of the threat is now widespread, however, emissions continue to surge, and serious global action to turn the tide is still nowhere to be seen. If we are to have any chance of preventing a bad situation becoming even worse, then we need to begin to seriously curtail emissions. Not in ten years' time, or even five, but now. Today.

As previously addressed, a quick win can come from rowing right back on methane emissions. Because of its short residence time in the atmosphere, halving emissions by 2030 could reduce planetary heating by a welcome 0.3°C by 2045. But we need to go further. Fossil fuel companies, responsible for leaking around half of all methane emitted by human activities, need to be made – by law – to clean up their act immediately. We can all help reduce emissions from the other big, non-natural source of methane by cutting way back on our meat feasting and thereby reducing the demand for livestock.

A sharp fall in methane emissions would be a great start to getting on top of the climate emergency, but much else needs doing in parallel. Fossil fuel corporations have to be brought to heel, and quickly; the wellheads and coal mines shut down as soon as possible. There is no way the fossil fuel sector is going do this voluntarily, so it must be compelled to leave known oil, gas and coal reserves in the ground, and to stop exploring for more. This can be accomplished via a combination of punitive measures, starting with the scrapping of subsidies. It is infuriating that the same world leaders who talked a good game at COP26 hand over wads of money to the fossil fuel sector in subsidies that, according to the International Monetary Fund, totalled $5.9 trillion in 2020 alone – that's $11 million *every minute*. There is no excuse for this, and it needs to stop now, or better still, the subsidies should be switched to the renewables sector.

In tandem, the issuing of new exploration licences must be brought to an end. Even as US President Joe Biden was telling the world that his administration was working over-time to 'show our climate commitment is action, not words', he was auctioning off more than 80 million acres of seabed in the Gulf of Mexico for fossil fuel exploration. This is bordering on the insane.

Alongside abandoning subsidies and ending the licensing of new exploration, banks must be made, using bouquets, brickbats or both, to stop hurling money at the fossil fuel sector. At the same time, unrelenting pressure must be applied to pension schemes, sovereign wealth funds, universities, churches and other major institutions to accelerate divestment of any and all stocks and shares they hold in fossil fuel companies. Insurers need to play their part by backing away from providing a safety net for fossil fuel facilities, such as rigs and refineries. Together, such a package of measures would go a long way towards making the extraction of oil, gas and coal uneconomical, especially as renewables continue to get cheaper. Throw in a carbon tax levied at the wellhead and mine entrance, and the fossil fuel industry could, and should, wither on the branch.

Restoring damaged and desecrated land can also play a huge part in limiting the global average temperature rise as much as possible. Before humans appeared on the scene, it is estimated that around 6 trillion trees populated the surface of our world. During the course of our civilisation's growth, however – more than half of these have been cleared, and the destruction shows little sign of stopping. Every year, 16 billion trees are grubbed up or chopped down, the great majority in the vast areas of ancient tropical forest that are being shorn at a rate of 476 trees every second. This not only has to stop but must be reversed.

One great and astonishing piece of news is that it looks as if tropical rainforests – if left to their own devices – can get back to almost 80 per cent of their old-growth condition in just twenty years, although more than a century is needed for them to return to their original, biologically diverse state. But right now, we need more trees – lots more. Large-scale tree planting, done right, unequivocally provides the best and cheapest way of pulling excess carbon out of the atmosphere. A 2019 study, for example, demonstrated that planting 1.2 trillion trees on a little more than 10 per cent of our planet's land surface – an area equal to China and the United States combined – could ultimately soak up almost one-third of all the carbon dioxide released by human activities that remains in the atmosphere today. There are still issues with reforestation, not least the fact that planting should not take cropland out of production, nor should it take the form of monocultural plantations with poor biodiversity. This said, it has a critical role to play in bringing down emissions.

An excellent measure to accompany widespread reforestation would be the progressive phasing out of beef and dairy farming. With beef and dairy accounting for nearly 15 per cent of all greenhouse gas emissions, the reforesting, where possible, of land currently used to rear beef and dairy cattle would provide a win–win situation. Together, just twenty of the world's biggest meat and dairy producers pump out more greenhouse gases than Germany or France, so one of the best things we can do, on a personal level, to help tackle the climate emergency, is to cut way back on our consumption of meat (especially beef) and dairy products.

And so the list of what needs to happen to bring emissions to heel goes on and on: restoring peatlands and wetlands so they store more carbon; cutting way back on flying and shipping consumer goods around the planet; launching massive investment programmes in home insulation and green domestic energy in developed countries. Electric cars are all well and good, but replacing the world's 1 billion personally owned fossil fuel powered vehicles with 1 billion electric ones brings its own baggage.

Most critically, this includes the energy, transport and environmental costs of large-scale extraction of lithium, nickel and other essential elements needed for battery construction. A far better way forward is via the ramping up of cheap, green and efficient public transport systems, journey-based carpools, car-share clubs and the incentivisation of walking and cycling. Equally importantly, industrialised nations need to start shovelling money in the direction of majority world countries to help them transition to greener futures with as little economic and social upheaval as possible – something they have so far been slow to do.

In order to limit the consequences of the climate chaos heading our way, the honest truth is that, of every decision taken, of every choice made – by individuals, local authorities, businesses big and small and governments – the question must be asked: is this good for the climate? If the answer is yes, all well and good. If the answer is no, then it cannot be allowed to proceed.

What might it be like to live in 2100?

The coming decade is very likely the most critical in human history. Inevitably, climate breakdown will hit us hard, but we can still cushion the blow substantially by taking urgent action now, both to cut emissions and to adapt to a much hotter world. We are at a fork in the road that one way leads to a calamitous and unsustainable future, and the other to a world in which rapidly falling emissions have slowed the rate of heating and large-scale adaptation has led to much greater resilience. The two perspectives of a future London that follow provide a flavour of what the choices we make in the next few years might mean for our world 80 or so years hence, and for those who live in it.

***

The late August bank holiday of 2100 dawns hot and clear, but there is no sign of Londoners attempting a quick getaway to reach the nearby beaches or the countryside. In fact, the streets of London are all but empty. Those lucky enough to have air conditioning, when the power is on, are cloistered in their homes from the scorching heat; the thermometer already touching 32°C and forecast to reach 44°C later in the day. Those without access to conditioned air have either joined the exodus north to escape the terrible summer heat, are camped out in the treeless and desiccated parks, or are sheltering in the shadows of the derelict streets. No one buys property in London any longer and those that own it can no longer sell.

In the south and east of the city, the sea has made deep inroads and shanties have grown up where a coastal breeze takes the edge off the heat and humidity, but there is a deadly price to pay in the form of endemic cholera and malaria. This is a killing ground for other reasons too, as the indigenous destitute and the penniless migrants who have fled even warmer Southern Europe and beyond scrabble violently to protect what little they have. In Westminster, the Houses of Parliament still stand, but the sound of debating has long gone, driven northwards to Carlisle by the smell of raw sewage flushed daily into the Thames by flash floods accompanying the evening thunderstorms.

To the west, just one runway and a single terminal remain open at Heathrow Airport. Since the swingeing personal carbon levies were imposed – a far too tardy response to try to slow climbing emissions – flying is a luxury only the very rich can afford. All around the city, the landscape is brown and parched, much of the ancient water distribution system long ago succumbing to settling and subsidence as the soils dried out, shattering mains and contaminating supplies. The motorways that head out from the city like the spokes of a wheel carry little traffic. Construction of the infrastructure to support electric vehicles never really got going and since the recent economic crisis, driven by the final collapse of the fossil fuel industry, the cost of fuel has been astronomical.

As the Sun climbs higher in the sky, a hot wind builds, ruffling the thick coat of dust and blown topsoil that blankets streets and buildings after six years of extreme drought. Carried aloft by air currents, it hangs like a grey cloud over the city but does little to ease the Sun's blazing light.

***

The late August bank holiday of 2100 dawns hot and clear, the Sun beating down on a city that glows emerald in its brilliant light. From the air, the London cityscape is a patchwork of green, white and silver, every roof either covered with greenery, painted white to reflect the Sun's rays or clad in solar panels. Streets are barely visible beneath spreading canopies of trees that keep the worst of the heat at bay. It looks like being another scorcher, but temperatures are down on previous summers, a trend that is forecast to continue.

The sounds of children playing in the traffic-free avenues are carried far and wide in the pristine, pollution-free air. Gaps in the foliage provide glimpses of chattering crowds, strolling on foot or riding bicycles, as they enjoy the leafy shade. Plentiful public transport powered by wind, solar and the new network of tidal barrages have made cars largely redundant in the city and all but the main thoroughfares have been reclaimed for the enjoyment of Londoners.

Streets are lined with homes that have been retrofitted to be carbon-free and insulated to keep out the summer heat. Deep channels below street level carry excess water from flash floods north and south to the river and east to where the sea has encroached inland. Here, wetlands host a huge variety of wildlife and provide all sorts of water sports for bank holiday visitors.

To the west, giant, fuel-cell-powered airships come and go from their Heathrow terminus, some slowly descending their mooring masts to offload, others detaching, turning like enormous white birds and heading off sedately to destinations at home and abroad. All around the city margins, smallholdings powered by solar grow the fruit and vegetables that land every day on the tables of city residents. Drought can still be a problem, but an efficient system of water recycling and catchments designed to make the most of rainfall from the evening storms mean that there is usually plenty to go around. The motorways that used to head out from the capital in all directions are mainly gone now, replaced by fast electric trains and tram networks or demolished and redeveloped as market gardens. Goods are brought to the city by rail and unloaded at giant distribution centres around the periphery for onward journeys in small electric trucks.

As the Sun climbs towards its zenith, people migrate towards the shaded street cafes and bars to enjoy a cooling drink away from the heat. The tinkling notes of a street piano drift up through the greenery and across the rooftops, mingling with the sound of animated conversation and the chatter of birds in the trees.

AFTERWORD

If you have made it this far, I suspect you may either be seething at what you perceive to be gratuitous alarmism or biting your nails with worry. To those who feel that what I have had to say is alarmist, I say this: in the sense of drawing attention to how bad things can get as the planet continues to heat up, I am certainly raising the alarm, and I don't apologise for this. But alarmist? No. There is no exaggeration of the dangers here, no hyperbole. All the material included and addressed in this book is rooted in hard science, underpinned by meticulous observation and careful modelling. Raising the alarm, in our current circumstances, is a good thing. It fits with the precautionary principle and also with the idea that we need to *really* know our enemy – in this case global heating – and how well it is armed, if we want to defeat it. My view is that, currently, most members of the public, and indeed most world leaders, simply do not. The fact that the word 'cake' was mentioned ten times more than 'climate change' on UK television in 2020 says it all about how true appreciation of the nature and scale of the climate emergency has yet to break through.

The truth is, playing down the potential worst effects of global heating and climate breakdown is far worse than raising the alarm and amounts to what I like to call climate appeasement. It does nothing to help spur the urgent action that is required, and by underplaying the climate threat it works – intentionally or not – to encourage a grudging and cautionary approach to emissions cuts that we can no longer afford.

On the other hand, if you have been worried or frightened by what you have read, that's good, you should be, especially on behalf of your children and their children. But don't let fear feed inertia. Fear does not have to be paralysing. Indeed, it is often the driver of effective action. No one ever won a war while knowing no fear, and make no mistake, this is a war. Wherever we live on this magnificent planet, we all need to do our utmost to try to keep it that way. The fact that the future looks dismal is not an excuse to do nothing, to imagine it's all too late. On the contrary, it is a call to arms.

So, if you feel the need to glue yourself to a motorway or blockade an oil refinery, then do it. In his book *How to Blow Up a Pipeline*, Andreas Malm argues convincingly that, such is the scale of the climate crisis, sabotage and property damage are absolutely justified in the battle against fossil fuel companies and others working against the public good. I understand that this is not to everyone's taste, but there is plenty more you can do. Drive an electric car or, even better, use public transport, walk or cycle; stop flying; switch to a green energy tariff; eat less meat; spread the word about the predicament we find ourselves in among your friends and family; lobby your elected representatives at both local and national level; and use your vote wisely to put in power a government that walks the talk on the climate emergency.

The problem is, though, even the greenest administration is not entirely free to ensure that the right measures are in place to kick global heating into the long grass. Within a political–economic system predicated upon competition and profit rather than the greater good, it is always going to be challenging and problematical – even with the best will in the world – to bring down emissions rapidly enough to avoid the most severe impacts of global heating. Many high-profile activists argue, then, that what's needed to tackle the climate emergency effectively is system change, and they are absolutely right.

Our climate is being destroyed by unadulterated, free-market capitalism – an ideology that simply cannot be sustained on a small planet with limited resources. It is a system that has no interest in the greater good and that rewards inordinate capital and the few that have it, rather than the majority who don't. It cares nothing for the environment or biodiversity and doesn't give a fig about the fate of future generations. In fact, it is *exactly* the wrong economic system to have in place at a time of global crisis. The bankruptcy of the system is especially well upheld in the grossly asymmetric partitioning of carbon emissions between the rich elite and everyone else.

One quick way of making a serious dent in emissions would be to take away what seems to be a free pass to pollute from the richest 1 per cent, who were responsible for 13 per cent of emissions in 2013. By 2030, this tiny elite is predicted to pump out 16 per cent of global emissions, 70 tonnes of carbon a year per person, when each of the poorest 50 per cent of the world's population – those who will bear the brunt of climate breakdown – are responsible for one measly tonne. For comparison, each UK citizen emitted 8.4 tonnes of carbon dioxide in 2021. Looked at individually, the annual carbon footprints of some of the world's mega-rich are staggering, uplifted to extraordinary levels – nearly 34,000 tonnes in one case – by their monstrous playthings: fleets of high-performance cars, homes on every continent, private jumbo jets, super-yachts and the like.

Emblematic of an economic system that is not fit for the purpose of transforming our society to one that matches the size of our world and its resources is the new rich person's toy – the spaceship. Weighing in at around 75 tonnes, the emissions expelled by a ten-minute flight on Branson's Virgin Galactic rocket are equivalent to an entire lifetime's emissions of one of the poorest billion people on Earth. At the height of a planetary emergency, this is plain wrong.

The measure of the maturity of any society must be how well it looks after the needs of every one of its people, and how it cares for the planet and all life thereon, by which metric we are little more than toddlers flailing about aimlessly in the dark. This will continue to be the case until the penny finally drops that we will never see off global heating without embracing a new way of doing things, which has nothing to do with the number of super-rich we can launch into space. There needs to be a sea change in the way economic success is measured, so that the accumulation of wealth is subordinated to how little carbon we emit or how much we manage to soak up.

Currently, the success of a national economy is measured in terms of its gross domestic product (GDP), which, in turn, is based purely upon the country's wealth. Other consider-ations, such as the health and well-being of the population, inequalities between the rich and poor, environmental issues and success in bringing down carbon emissions, form no part of it.

Although not perfect, one way of changing this to help tackle the climate and ecological emergency – which has been flagged by enlightened economists for some time – would be via a switch to a so-called quality adjusted GDP metric. Under this metric, good things, such as carbon-reduction measures, are rewarded, while bad stuff, for example, those products or services linked to high emissions or which are environmentally damaging, are marked down. A GDP metric that operated along these lines would benefit everyone by linking the money in people's pockets to national and global indicators of progress on emissions reductions and ecological improvements. Transitioning to such a system would not be easy, but it can be done provided the will is there. Without it, making serious inroads into the dangerous and growing levels of carbon in the atmosphere is likely to be all but impossible.

At the time I wrote this, on the last day of 2021, the UK was basking in record-shattering late December warmth, with the temperature climbing to 16.5°C in Bala, north Wales. A few days ago, on the other side of the world, the mercury touched 19.4°C in Alaska, compared to an average December daily mean temperature of around zero. In some of the Italian ski resorts it is too warm – at an extraordinary 15°C – for even artificial snow. Meanwhile, high up in the Colorado Rockies, unheard-of winter wildfires in the past 48 hours have destroyed more than 900 buildings and forced thousands from their homes.

As I tie up loose ends in March 2022, eastern Australia has just experienced record rainfall, driving some of the worst ever flooding across the region, while more than 700 wild-fires are raging across Texas. In Eastern Europe, meanwhile, the continuing conflict in Ukraine provides a tiny foretaste of the migration problems and commodity shortages to come, as climate breakdown ramps up. For many, as we head further into 2022, it is already a different world out there. Soon it will be unrecognisable to every one of us.

We may no longer be able to give dangerous climate breakdown the slip, but we still have the means to fend off a climate cataclysm that may threaten the very survival of human civilisation. In the decades since the first UN COP Climate Change Conference in 1995, we have used up an entire bale in prevarication and inertia, so all we are left to clutch at is the last straw. We cannot fail to grasp it.

ONLINE AND OFFLINE RESOURCES

Online

Climate Action Tracker
Keep on top of how your country is doing in cutting emissions:
https://climateactiontracker.org

Climate Central
Excellent climate research and news:
https://www.climatecentral.org

Cool Earth
Bill McGuire's opinion and comment blog focusing
primarily on global heating and climate breakdown:
https://billmcguire.substack.com

Extinction Rebellion
Do your bit to help tackle the climate emergency:
https://rebellion.global

Intergovernmental Panel on Climate Change
Climate breakdown science from the horse's mouth:
https://www.ipcc.ch

NASA Global Climate Change
Stay up to date on the key global heating and climate breakdown
metrics: https://climate.nasa.gov

RealClimate
Climate science from climate scientists:
https://www.realclimate.org

Skeptical Science
Getting sceptical about global warming scepticism:
https://skepticalscience.com

Met Office UK Climate
Everything you need to know about the UK climate
and how it is breaking down:
https://www.metoffice.gov.uk/weather/climate/uk-climate

Offline

Berners-Lee, Mike, *How Bad are Bananas? The Carbon Footprint of Everything* (London: Profile Books, 2020)

Earle, Steven, *A Brief History of the Earth's Climate: Everyone's Guide to the Science of Climate Change* (Gabriola, BC: New Society Publishers, 2021)

Hansen, James, *Storms of My Grandchildren: The Truth About the Coming Climate Catastrophe and Our Last Chance to Save Humanity* (New York: Bloomsbury, 2011)

Klein, Naomi, *This Changes Everything: Capitalism vs the Climate* (London: Allen Lane, 2014)

Malm, Andreas, *How to Blow Up a Pipeline: Learning to Fight in a World on Fire* (London: Verso Books, 2021)

McGuire, Bill, *Skyseed* (Market Harborough: The Book Guild, 2020)

Mann, Michael, *The New Climate War: The Fight to Take Back Our Planet* (Brunswick, Victoria: Scribe Publications, 2021)

Wallace-Wells, David, *The Uninhabitable Earth: A Story of the Future* (London: Allen Lane, 2019)

INDEX